D1866738

ALLE ZEIT WACH
SJ
1842

Psychophysics in Action

Edited by

Gunilla Ljunggren and Stan Dornic

With Contributions by

O. Bar-Or · A.-M. Bonnel · H. Eisler · Å. Hellström
D. P. Hunt · W. A. Latzka · L. Levine · H. Linderholm
L. E. Marks · K. F. Metz · S. R. Muza · K. B. Pandolf
C.-A. Possamaï · R. J. Robertson · P. T. Sagal · M. R. Sams
B. Scharf · J. C. Stevens · R. Teghtsoonian

With 38 Figures

Springer-Verlag
Berlin Heidelberg New York
London Paris Tokyo

Dr. Gunilla Ljunggren
Dr. Stan Dornic
Department of Psychology
University of Stockholm
106 91 Stockholm, Sweden

ISBN 3-540-50686-1 Springer-Verlag Berlin Heidelberg New York
ISBN 0-387-50686-1 Springer-Verlag New York Berlin Heidelberg

Library of Congress Cataloging-in-Publication Data. Psychophysics in action. In honor of Gunnar A. V. Borg on the occasion of his 60th birthday. Includes bibliographies and index. 1. Psychophysics. 2. Borg, Gunnar. I. Ljunggren, Gunilla, 1943– . II. Dornic, Stanislav. III. Bar-Or, Oded. IV. Borg, Gunnar. [DNLM: 1. Psychophysics. WL 702 P9745]
BF237.P88 1989 150.19'8 89-6079
ISBN 0-387-50686-1 (U.S.)

This work is subject to copyright. All rights are reserved, whether the whole or part of the material is concerned, specifically the rights of translation, reprinting, re-use of illustrations, broadcasting, reproduction by photocopying machine or similar means, and storage in data banks. Duplication of this publication or parts thereof is only permitted under the provisions of the German Copyright Law of September 9, 1965, in its version of June 24, 1985, and a copyright fee must always be paid. Violations fall under the prosecution act of the German Copyright Law.

© Springer-Verlag Berlin Heidelberg 1989
Printed in Germany

The use of registered names, trademarks, etc. in this publication does not imply, even in the absence of a specific statement, that such names are exempt from the relevant protective laws and regulations and therefore free for general use.

Product Liability: The publisher can give no guarantee for information about drug dosage and application thereof contained in this book. In every individual case the respective user must check its accuracy by consulting other pharmaceutical literature.

Typesetting, printing and bookbinding: Konrad Triltsch, Graphischer Betrieb, Würzburg
2126/3130-543210. Printed on acid-free paper

In honor of

Gunnar A. V. Borg

on the occasion of his 60th birthday

Preface

Gunnar A.V. Borg was born in Stockholm on 28. November 1927. Educated at Stockholm University, he obtained his Ph.D. from the University of Lund in 1962. Subsequently he held various teaching and research appointments at the University of Umeå in northern Sweden, where he also served as President of the Graduate School of Social Work and Public Administration in 1966–1967. In 1971 he was appointed Professor at Stockholm University, where he headed the Institute of Applied Psychology for over a decade. Since 1980 he has been at Stockholm University's Department of Psychology, and in 1987 he received a Professorship in Perception and Psychophysics. Over the last 20 years he has held several visiting appointments abroad, particularly in the USA, and has lectured at many universities both in and outside Europe.

From the beginning of Gunnar's research career, his thinking has been affected by Gestalt psychology as well as by some principles of theoretical philosophy. The former has not only influenced Gunnar's early unconventional works on "gestalt strength" but also one of his major areas of thinking, the notion of "total perceived effort," which can be seen as a "gestalt" underlain by a variety of sensory data. The philosophical influence is obvious: Gunnar's papers reveal an abiding concern with epistemological issues, issues that are linked to his persistent attempts at making interindividual comparisons. These attempts resulted in his now renowned range model, which has rendered individual differences in human perceptual functioning more understandable. According to this principle, maximum perceptual intensity is set subjectively equal over individuals, and the intensity of a submaximal perception (any perception) is determined by its relative position in the range.

One of the pivotal notions that is ubiquitous in Gunnar's research – concerning perceived effort or perceived exertion – emerged in the 1950s and can be seen to reflect Gunnar's profound and persistent human and scientific interest in the covert cost of man's overt performance.

In the following years, Gunnar explored the entangled relations between words and numbers – a temporary trip into quantitative semantics that later helped him in the construction of his now famous RPE (rating of perceived exertion) scale. Far from being one-sidedly interested in academic topics, he also designed a bicycle ergometer that made it possible to make effective use of psychophysical methods in the study of physical work. Since the mid-1960s Gunnar's name has been associated with a unique, widely quoted experiment that demonstrated close agreement between subjective responses (psychophysical ratings) and responses from afferent nerves.

Towards the end of the 1960s, when the notion of perceived exertion had already been well anchored in a number of laboratory experiments, it began to make its way into "real-life" domains – such as clinical, ergonomic, and sports areas.

For instance, Gunnar's psychophysical methods proved highly useful tools in assessing clinical symptoms, physical strain, and thus even work capacity in different patient groups. Later on, his approach was successfully applied in analyzing the notion of difficulty, ubiquitous in mental work, which had earlier been conceived of only in terms of performance.

The RPE scale, which was constructed such that it could be used to make direct measurements of subjective levels in clinical and ergonomical contexts where orthodox psychophysical ratio scaling was not appropriate, saw increasing use and gained international recognition during the 1970s. At the same time, Gunnar was widening the scope of his approach, theoretically advancing and modifying Stevens' general power law and further developing scaling methodology; a comparatively recent example is his CR (category-ratio) scale, – a category scale with ratio properties that significantly enhances the sensitivity of psychophysical measurements.

During the last few years, Gunnar has been paying increased attention to his range model, which might be perhaps better thought of as a "range principle," a "range perspective," or even a "range approach" to human functioning. This approach goes far beyond being a methodological tool in psychology. The "range thinking," with a potentially strong human impact on the study of man, is touched upon by a prominent American philosopher in the first chapter of this book.

Gunnar's research has many unconventional features. Some of his ideas ran contrary to then established streams of thought, and sometimes he found himself engaged in an uphill fight. It takes strong belief in one's own work not to give up. He never gave up – to the advantage of a realistic and differentiated view of man in modern psychology.

One of the most salient features of Gunnar's research is that it spans seemingly distant and sometimes seemingly incompatible areas, from the purely theoretical to the markedly applied. It has been fruitful in advancing disciplines as different as psychology, clinical medicine, and ergonomy.

Like Gunnar's own research, the articles in this volume span a wide range of topics. The book's introductory chapter confronts the reader with a set of central philosophical issues concerning the mind. Subsequent chapters divide into two main groups. The first group of chapters engages the reader in various challenging theoretical and experimental issues: modeling, discrimination, performance and learning, cross-modal perception, attention, psychophysical similarity, and individual differences. The second group stimulates the reader with topics centering around physiological and clinical issues in the measurement of perceived exertion: clinical assessment in the very young, studies of exertion and effort during cycling, medical diagnostic exercise testing, and physiological determinants. The multiplicity of topics, represented by contributions from many prominent authors who kindly agreed to collaborate on the book, reflects Gunnar's own many different interests; at the same time, this very diversity illustrates the vast and rich arena of modern psychophysics in action.

GUNILLA LJUNGGREN
STAN DORNIC

Contents

Physiological and Clinical Aspects

List of Contributors

Bar-Or, O., Children's Exercise and Nutrition Centre, McMaster University, Chedoke Hospital Division, Hamilton, Ontario, Canada

Bonnel, A.-M., Unité de Neurosciences Cognitives, CNRS, Marseille, France

Eisler, H., Department of Psychology, University of Stockholm, Stockholm, Sweden

Hellström, Å., Department of Psychology, University of Stockholm, Stockholm, Sweden

Hunt, D. P., Experimental Training Center, Department of Psychology, New Mexico State University, Las Cruces, NM, USA

Latzka, W. A., U. S. Army Research Institute of Environmental Medicine, Natick, MA, USA

Levine, L., U. S. Army Research Institute of Environmental Medicine, Natick, MA, USA

Linderholm, H., Department of Clinical Physiology, University of Umeå, Umeå, Sweden

Marks, L. E., John B. Pierce Foundation Laboratory and Yale University, New Haven, CT, USA

Metz, K. F., Human Energy Research Laboratory, Program in Health and Physical Education, University of Pittsburgh, Pittsburgh, PA, USA

Muza, S. R., U. S. Army Research Institute of Environmental Medicine, Natick, MA, USA

Pandolf, K. B., U. S. Army Institute of Environmental Medicine, Natick, MA, USA

Possamaï, C.-A., Unité de Neurosciences Cognitives, CNRS, Marseille, France

Robertson, R. J., Human Energy Research Laboratory, Program in Health and Physical Education, University of Pittsburgh, Pittsburgh, PA, USA

Sagal, P. T., Department of Philosophy, New Mexico State University, Las Cruces, NM, USA

Sams, M. R., Experimental Training Center, Department of Psychology, New Mexico State University, Las Cruces, NM, USA

SCHARF, B., Auditory Perception Laboratory, Northeastern University, Boston, MA, USA

STEVENS, J. C., John B. Pierce Foundation Laboratory and Yale University, New Haven, CT, USA

TEGHTSOONIAN, R., Department of Psychology, Smith College, Northampton, MA, USA

Introductory Philosophical Considerations

Borg's Postulate and the Philosophy of Mind

Paul T. Sagal

Abstract

In a previous paper the author and Professor Borg examined the relevance of Borg's range principle to the philosophical problem of other minds. The present paper provides further elaboration and analysis. It also attempts to confront the so-called mind–body problem and raises the question of what philosophical views on the mind–body problem are compatible with Borg's empirical work in psychophysics. Also included is a discussion of the "inverted spectrum" argument and its relationship to Borg's work.

The problem of other minds, like many philosophical problems, pulls us strongly in opposing directions. Of course, I can know that you have a headache. I offer you aspirin, I express concern, I empathize with you. You may, of course, be faking, but where I can see no motivation for this, surely I can be certain that you have a headache. How do I know? I see you grimacing, clutching your head; you tell me of your pain. I don't really have the slightest doubt. On the other hand, and with philosophical problems there is always an "on the other hand", maybe I don't really know you have a headache. You *could* be faking even if I can't determine any motivation for such faking. Really, all I have to go on is your behavior, including your verbal reports. I infer the inner mental state. What guarantees the correctness of the inference? I do know that in my own case headache behavior is usually correlated with having a headache. But how do I know that you are like me in this respect? You appear to be like me, you are a person just the way I am. But how do I know this? Maybe you are a robot, or maybe not all human beings have conscious experiences in the way I have. The trouble seems to be that I can only experience my own experiences. We are pulled in two opposing directions.

There have, of course, been many attempts to solve or dissolve the problem. There seems to be some ambiguity in the concept of knowledge. The skeptical arguments apparently employ a stricter standard of knowledge than does the common sense argument. Is it fair to require the kind of certainty the skeptic is demanding? Of course, I can't have your headache. If I did, it wouldn't be your headache. Isn't the skeptic asking for the impossible? On the other hand, just because we are usually quite comfortable in saying we know that other people have experiences like headaches doesn't by itself entail that we really do have such knowledge.

One approach suggests that we adopt some compromise position on the nature of knowledge between the excessively strict conception of the skeptic and the excessively liberal conception of the defender of common sense. A scientific conception is just such an in-between position. The general principle that others have minds in much the same way that I have and that in many circumstances I can know that others have headaches provides the best explanation of observable behavior. This is the approach that Ziff (1965) and Putnam (1972) have adopted and the one that will provide the framework for this paper. One advantage of the scientific approach is that it allows scientific considerations to contribute to the resolution of a traditional philosophical problem. This is where the psychophysical research of Gunnar Borg into problems of interpersonal comparisons of experience enters the picture.

I first met Professor Borg at a small social gathering at New Mexico State University where I am a Professor of Philosophy and Professor Borg was a visiting Professor of Psychology. We got to talking about Professor Borg's research and it immediately struck me that we both shared a common interest in the problem of other minds. Of course, I was not familiar with the empirical work of Professor Borg and he did not have a clear idea of the philosophical consequences of his work. He was, however, aware of the possible relevance of this work to philosophical concerns and was pleased with my suggestion that we try to collaborate on a paper that would underline the philosophical relevance of his work, especially the so-called *range principle* which holds that the range of experience of different subjects is the same. Roughly, Borg held that if we accept this principle then we should be able to compare the experiences of different people in an empirically testable way. That summer, I visited Professor Borg in Stockholm and we ultimately produced a paper entitled "Borg's Postulate and the Problem of Other Minds."

What exactly is the contribution of Borg's work to the problem of other minds? Its place lies within the context of the scientific approach. Let us suppose that the other minds framework provides the best scientific explanation we have available for the relevant observable behavior; we still need to develop the framework in such a way as to yield empirically testable and hopefully verifiable predictions. In other words, let us say that Herby and Harvey are both engaged in strenuous physical activities. They are both exerting effort and are experiencing the effort expended. Both say they are working very hard but not at their maximum. Let us say they characterize their effort and their experience of effort as about ¾ of their maximum. We want to be able to compare their experiences. First of all, we assume that they are experiencing something (this is part of the assumed other minds hypothesis). But what are they experiencing? Can we make empirically testable judgements with respect to their respective experiences? This is where Borg's range principle comes in. We assume that they both have the same range of experience. We establish these individual ranges of experience of effort independently. We correlate experience with physically determinable qualitities, e.g., heartbeat, amount of lactate in the blood. We thus establish what philosophers of science call a bridge principle, or a rule of correspondence. This relates unobservable (theoretical) phenomena to observable phenomena. Here of course the unobservable phenomena, at least for the scientific investigator, are the individual's experiences of effort. Borg's research shows us that the appropriate predictions turn out to be verified. Thus Borg's postulate helps us to obtain scientific knowledge of other minds. It helps to explain how interpersonal comparisons of experience are possible. The postulate is itself a part of a theory which enables us, by using relevant bridge principles, to verify predictions about the experiences of others and how these compare with one another.

Borg's whole approach seems contrary to a tradition in psychology and psychophysics which holds that experience is not suitable subject matter for a science of psychology, and that even if psychology could deal with experience, it would still be impossible to make interindividual comparisons of experience. S. S. Stevens has been a strong proponent of the latter view, a view which seems to have its roots in the skeptical philosophical position outlined earlier. Borg's work by itself does not, and cannot, refute Stevens' skepticism. What it does help us with is the problem of how we can compare individual experiences given certain assumptions about the existence of such experiences, and how we can relate such experiences with observable phenomena. Borg's range postulate fits in with this framework. It also shows us how we can incorporate verbal reports of subjective experience into a scientific framework for intersubjective comparison of experience.

Following Ziff (1965) and Putnam (1972), we have been speaking of other minds as providing the best explanation of the behavior of others. We need to be more explicit about just what this entails. First of all, what is mind in the sense of the other minds explanation? We are not committed to regarding the mind as an entity distinct from the body, we are not committed to Cartesianism (the ghost in the machine). We are committed to the existence of mental states or functions like thinking, believing, wanting, and choosing. An individual has a mind only if it thinks, believes, wants, chooses, etc. These may be Cartesian minds, but this does not follow from the existence of mind in the sense of the other minds explanation. The mind we have may be a Cartesian mind, but "having a mind" might simply be a way of speaking, a way of abbreviating talk about the mental states and functions we have been discussing. We can see how the other minds problem leads to the *mind–body problem,* the problem of the nature of mind, and the relation of the mental to the physical. There is, I think, only one view on the mind–body problem that is excluded by the other minds explanation and this is the view there is no mind in the sense that there are no mental states and functions. This view is sometimes termed eliminative materialism. Quine (1985) is perhaps its most prominent contemporary spokesman. On this view, we can eliminate mental states by speaking instead of the physical states (neural states) which correspond to them. Eliminating mental states is eliminating mind in our sense. The other minds explanation is replaced by the other brains explanation. Such an explanation, or explanatory framework, does offer some advantages over the other minds one. The principal advantage is that brain states are physical entities and our biology and neurophysiology are already committed to their existence. Since we already have these entities, other things being equal, it would serve our needs for ontological economy to make use of them for psychological explanation.

So do we not have a better explanation of behavior than other minds? I think not, for one very simple reason. The other brains explanation is false, i.e., the theory which provides the explanation is false. How so? I know that I have mental states. I know this directly, through introspection if you will, so a theory which holds that there are no mental states cannot be correct. But maybe we have taken this talk of elimination too seriously. Maybe all that is changed is that we do not need other minds to explain behavior even if we grant that there are other minds. Can we explain behavior, however, without talking of mental states and functions? This depends upon what we mean by explanation. The explanation we are seeking is a psychological explanation, it is an answer to a question in terms of belief, desires, etc. Explanation in strictly physical terms will only do if in some sense phsychological talk is reducible to this other talk. But this must be accomplished without eliminating the psychological talk altogether, for this would change the very concept of explanation we are employing. This seems to lead us to the position that mental states may be identical to physical states, the identity theory. If this is true, we seem to have the best of both worlds: (1) we keep mental states and functions and along with them psychological explanations, but (2) we do not expand our scientific ontology beyond the physical. This view does seem to me to be compatible with the other minds explanation; but I am not certain because I am not sure I understand the identity theory. Identity is after all a symmetrical relation. The identity theory then does not by itself reduce the mental to the physical any more than it reduces the physical to the mental. And if reduction is in the background, if the identity is to be construed as a reduction, then I don't see any difference between the identity theory and the elimination theory. Quine sees no difference between identifying the mental with neural states and eliminating mental states in favor of neural states. How can Quine then obtain what amounts to a reductionist reading of identity statements? The answer seems to be something like this. The system of science gets its meaning from the observation sentences the system entails. Neural states are already part of the system and already play a role in the derivation of observation sentences. Thus no new states need be added to get us from psychological explanations to relevant observation sentences. It doesn't make any difference to the derivation of observation sentences whether we speak of mental events or neural events since they are identical. But neural states get a privileged place because they are already integrated into the scientific system.

Is Quine's identity-cum-elimination theory compatible with the other minds explanation? I am not sure. Given Quine's views on empirical meaning, ontological commitment, reduction, and relativity, I am not sure the question makes sense. When I raised the question, I thought I was asking a relatively straightforward one. If we take Quine seriously, then I wasn't asking such a straightforward question at all. The answer I want to give is that Quine's position is not compatible with the other minds explanation, but I am not sure I can support this.

I should, following Putnam (1972), emphasize that other minds is best thought of as a framework for psychological explanation in much the same way as physical objects provide a framework for physical explanation. Neither minds nor physical objects by themselves explain anything. It is rather the configurations of particular mental states and functions, like the configurations of physical objects, which play a role in explanation. To the extent that we provide successful explanations, the explanatory framework receives support.

We began with the problem of other minds and showed where Borg's work on the range principle fits in. We then turned to the problems concerning the nature of mind in the attempt to explicate the other minds explanatory framework. I would like at this point to suggest to Professor Borg some additional questions for which an interdisciplinary approach (philosophy and psychophysics) might prove fruitful. I plan to do my best to keep him in touch with philosophers and philosophizing. This should not be too difficult since it is clear to me that Professor Borg has always had a sweet tooth for philosophy anyhow.

The key questions I would like to investigate with Professor Borg's help are those relating to the qualitative features of experience. Herby and Harvey both look at the traffic light and report that the light is green. How can I know that Herby and Harvey actually see the same color? That they are having some visual experiences is part of the other minds explanatory framework, but granting this, are there any assumptions we can reasonably make that would allow us to penetrate the qualitative features of their experiences? This is the old problem of the *inverted spectrum* – Herby sees green when and only when Harvey sees red, and their other color perceptions are also systematically inverted. Borg's range principle cannot help us here because presumably their ranges could be the same even with inversion. We can, it seems to me, say things about their wiring, their neurophysiological structures, and that identical structures should produce identical experiences, but how can we defend such a principle? Borg's principle enabled us to compare the experiences of effort of Herby and Harvey, but not to the extent of getting at the qualitative "feel" of the experience. Their experience could be said to be the same, but it is not clear that this sense of sameness entails sameness of qualitative content. We think we do know how effort is experienced in others on the basis of how effort is experienced by us. But this kind of an argument from analogy based on a single case seems incredibly weak, though we do, I think, actually reason in this way.

Is there a way to strengthen this argument from analogy? I am not certain Professor Borg would worry too much about this kind of question. It looks to be beyond the reach of scientific investigation. Reichenbach (1951) thought any answers to the question to be empirically unverifiable and hence (cognitively) meaningless and Professor Borg might have some sympathy with this, though the reader will call that S. S. Stevens held that psychophysics could not deal with interpersonal comparisons of experience and Borg rejected this view.

As Professor Borg and I walked through the woods near his home in Stockholm, we discussed some of the more speculative questions, especially, as I remember, the nature of consciousness. I did not get the feeling that Professor Borg wanted to commit himself about consciousness. He suggested that his work did not really commit him to specific views on the matter. I let the matter drop, but I would like to explore such matters with him further at a later date. Perhaps the present modest paper can serve as a stimulus.

References

Putnam, H. (1972). Other minds. In: Rudner, R. (Ed.) *Logic and arts: Essays in honour of Nelson Goodman.* Indianapolis: Bobbs-Merril.

Quine, W. V. O. (1985). States of mind. *Journal of Philosophy, 82.*

Reichenbach, H. (1951). The rise of scientific philosophy. Berkeley: University of California Press.

Ziff, P. (1965). The simplicity of other minds. *Journal of Philosophy, 20,* 575–584.

Theoretical and Experimental Aspects

Data-Equivalent Models in Psychophysics: Examples and Reflections *

Hannes Eisler

Abstract

Data-equivalent models are different (mathematical) models that make identical predictions. Conversely, different representations of the same data constitute data-equivalent models. A number of examples from psychophysics and multidimensional analysis are given, together with ways to choose between such models.

Introduction

One of the aims of science is to condense the variety of nature, or of phenomena within nature, in such a way that they become amenable to our understanding and make predictions possible. This is achieved by theory or by modeling phenomena in which the researcher is interested. The distinction between theory and model is fluid; usually a model is thought of as describing a more limited phenomenon, whereas a theory is assumed to be of wider scope, often comprising a number of models.

Here we shall mostly consider mathematical models, that is, models mapping natural phenomena into mathematical symbolism, thereby making quantitative predictions possible. Very often, different models are presented as descriptions of the same phenomenon, and a decision as to the most correct one is based on goodness of fit, though other criteria like the fit to some theory might be adduced. (It is not uncommon, however for a researcher to content himself with presenting an acceptable fit of his pet model, neglecting the existence of competing models.)

One example of a comparative study of models is to be found in Eisler (1981 a). In this study, duration discrimination data collected by Allan (1977) were analyzed. Seven models, in combination with five statistical distributions of subjective duration, were compared in terms of goodness of fit. One, the parallel-clock model (see pp. 12–15) with a normal distribution, proved superior, but even for this combination the fit was unsatisfactory. An improvement

---

* This work was supported by the Swedish Council for Research in the Humanities and Social Sciences.

of the model, introducing breaks in the psychophysical power function (again, see pp. 12–15), eventually resulted in an acceptable fit. This study illustrates the way the choice between different models is usually made.

In principle, the choice in such cases is easy. As the models make different predictions, for errorless data one model (at most) would be correct. Quite another problem is posed by data-equivalent models, that is, different models making identical predictions. A famous instance in physics is the restrictions required by quantum mechanics as to possible regions for electrons in atoms or molecules. These were modeled in 1925 by Heisenberg by means of matrix calculus and in 1926 by Schrödinger's wave equation, a differential equation. The two models yielded the same results, which agreed with empirical findings.

There is another, trivial, way in which models can be data-equivalent. The variables entering an equation can be transformed, as well as the mathematical operations of the equation, in such a way that the equation still holds. For instance, if we were to interpret the concatenation operation of placing two weights onto one pan of a balance, thereby equating another weight on the other pan, as multiplication rather than addition, the derived weight scale would differ from the common one with kilograms or pounds as units. (Taking logs of these scale values would bring us back to addition.) The weights, however, keep their equilibrium irrespective of how the concatenation operation is interpreted. The implication is that the choice of scale is, in a sense, arbitrary. Sometimes a new scale may be obtained by adding an additional parameter, as for instance when, in the Bradley-Terry-Luce model, scale values v are replaced by v^k (see Suppes & Zinnes, 1963, and their comment on p. 54. The Bradley-Terry-Luce model is similar to Thurstone's pair comparison model; see pp. 15–18. It differs from it by assuming the logistic instead of the normal distribution.)

Criteria other than the correctness of a model ("law", "invariance") have to be invoked for such cases of data-equivalence, for instance, simplicity of measurement theory, simplicity of substantive theory, communicability (see Eisler, 1965 b). The lesson to be learnt is that in talking about data-equivalent models we have to ask whether the equivalence is of this trivial kind or not.

Examples of Apparent and Real Data-Equivalence

Apparent Data-Equivalence in Subjective Duration

The parallel-clock model (Eisler, 1975 a, 1981 b) describes the behavior of an observer who is required to produce a duration as long (or half or twice as long) as a preceding standard duration. The model assumes two sensory registers. Subjective duration begins to accumulate in the first register at the onset of the standard, and in the second at the onset of reproduction. The two durations are experienced as equally long when the difference between the contents of the two registers (total duration and reproduction duration,

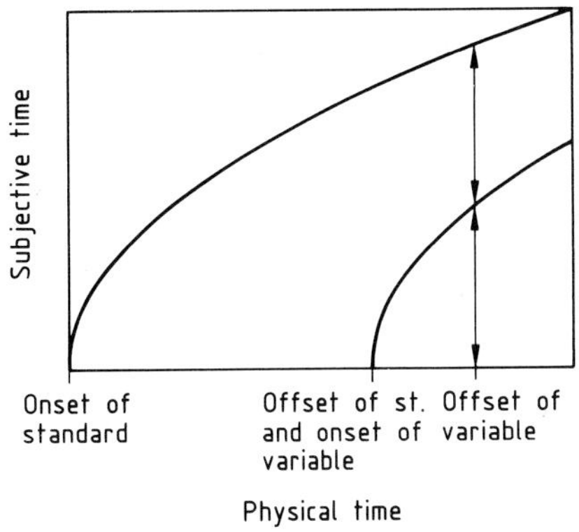

Fig. 1. Duration reproduction according to the parallel-clock model. Subjective versus total physical duration (*left curve*) and versus reproduction duration (variable; *right curve*). When the difference between these two subjective durations (*upper arrow*) equals the subjective reproduction duration (*lower arrow*), the observer reports equality between standard and reproduction by shutting off the sound

respectively) equals the content of the second register (reproduction duration). The observer then terminates the ongoing second duration (the durations are often indicated by a sound) by, say, a button press (see Fig. 1). In agreement with the model, when the reproduced durations are plotted against the total durations, i.e., the sum of the standard and reproduction durations (see, for example, Fig. 2), the plot is linear. From the slope and intercept two of the parameters of the psychophysical power function $\Psi = \alpha(\Phi - \Phi_0)^\beta$, namely, the subjective zero Φ_0 and the exponent β, can be computed (Ψ denotes subjective and Φ clock duration). For details, see Eisler (1974, 1975a).

However, most often a discontinuity or break is found in the psychophysical function for duration (likewise in scales obtained by magnitude estimation; see Eisler 1975a), implying different scale units α that compress or

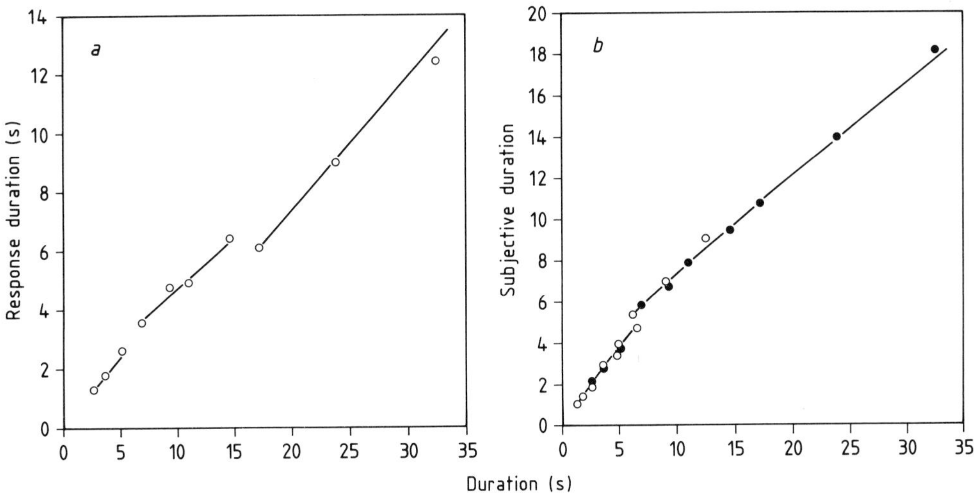

Fig. 2. a Reproduction durations plotted against total duration. **b** The psychophysical function: *filled circles* standard durations; *open circles,* reproduction durations

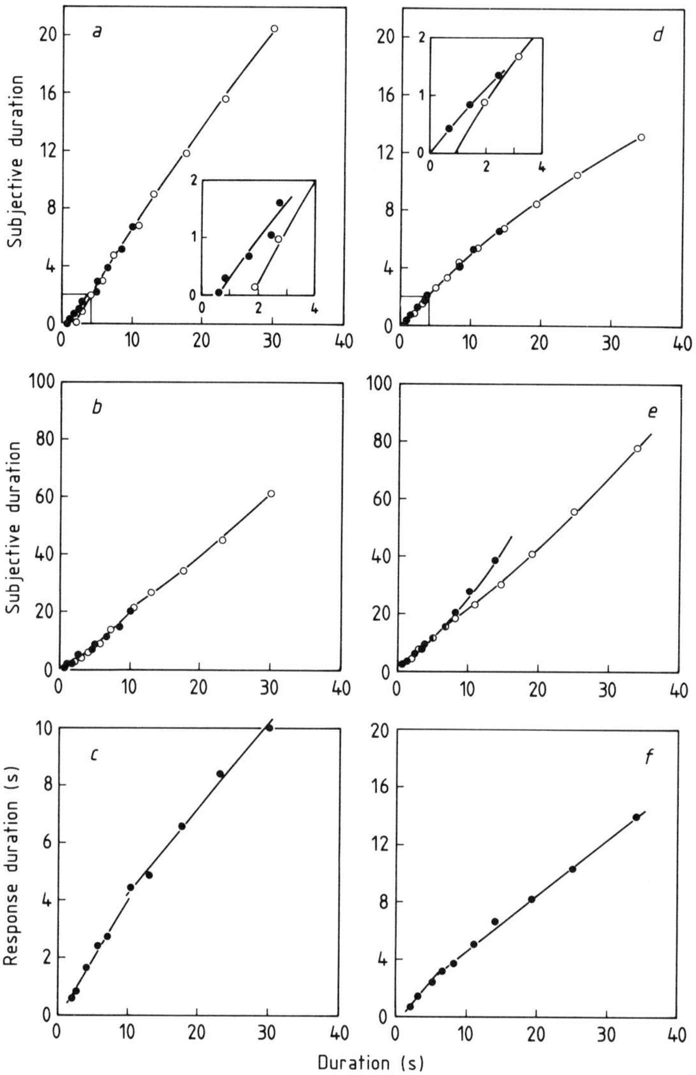

Fig. 3a–f. Duration halving (**a–c**) and reproduction (**d–f**). **c, f** The raw data; **a, b** and **d, e** the psychophysical functions according to two interpretations. *Insets* in **a, d** show the area near the origin. For details see text. (Data from Eisler, 1975a)

expand portions of the experimental range relative to each other, and/or different values of Φ_0. A consequence of such a break is that the plot of reproduction duration against total duration obtained from the raw data consists of three straight lines: For the outer two, reproduced and total duration lie on the same segment of the psychophysical function, the lower or the upper, respectively, whereas for the middle line, the total duration lies on the upper, and the reproduced duration on the lower segment. Figure 2 shows an example.

When the data, as in Fig. 2, consist of all three lines, there is no problem. All the parameters can be computed, the exponent β from the slope of either of the outer lines (which are parallel), the two values of Φ_0 from the intercepts of the outer lines, and the ratio of the two α values from the slope of the middle (transition) line.

Quite often, however, the data show only two of the three lines. Then, we do not know which one corresponds to the middle (transition) line, and which is the upper (or lower) line. Accordingly, two interpretations of the data at hand are possible: data-equivalence. If we regard the uppermost of the two lines as the transition line, then the lower line gives us the values of β and Φ_0, and the upper line the ratio of the two values of α. The opposite would be the case if the lowermost line were regarded as the transition line. Figure 3 gives an example.

Raw data from one observer are shown in Fig. 3c, f; Fig. 3c plots data from a halving experiment and Fig. 3f data from an equal-setting (reproduction) experiment. Treating the lower, left-hand lines as transition lines, we obtain psychophysical functions with an exponent below one, as shown in Fig. 3a, d. If we treat the upper, right-hand lines as transition lines, the exponents become larger than one, as in Fig. 3b, e. The curves in Fig. 3a, b, d, e are the corresponding fitted psychophysical functions.

I call this kind of data-equivalence *apparent*, because choosing between the two functions is rather easy. One way would be to repeat the experiment with a larger range of standards, which would probably yield a third straight line. Another way would be to compare the exponent obtained from this observer for these two tasks with the exponents from other tasks (magnitude estimation and doubling) collected in the same series of experiments for the same observer, as well as with exponents for other observers. As almost all exponents are below one, the psychophysical functions in the uppermost panels are chosen.

Data-Equivalent Models for Pair Comparisons

In a pair comparison task an observer (or a group of observers) is repeatedly presented with all possible pairs of a set of stimuli and required to indicate, for every pair, which stimulus is the stronger, the larger, the preferred one, etc. The range of stimuli in the set is sufficiently restricted that the probability of choosing one stimulus above another is never zero or one for any pair. The raw data from such an experiment constitute a matrix with the relative frequencies of choosing one stimulus above another as elements, the response matrix. According to Thurstone's (1927) case V, a subjective scale for the percepts corresponding to the stimuli can be defined if:

1. The observer's ordinal judgment is interpreted as based on the *difference* between the two percept magnitudes. (Thurstone seems to have taken this interpretation for granted, because it is not made explicit.)

2. It is assumed that the standard deviations of all percept magnitudes are equal.
3. It is assumed that the distributions of all percept magnitudes, and thus of all their pairwise differences, are normal.

Making other assumptions that are irrelevant to the present argument, and replacing the relative frequencies of the response matrix by the corresponding z values, the scale values of the percepts on an interval scale can be computed, and the consistency of the obtained scale tested.

However, a data-equivalent logarithmic interval scale can be defined by changing the conditions mentioned above to the following (Eisler, 1965 a):
1. The observer's ordinal judgment is interpreted as based on the *ratio* between the two percept magnitudes.
2. It is assumed that the standard deviations of all percept magnitudes follow Ekman's law (Ekman, 1959; Stevens, 1975, pp. 234–235), that is, that they are proportional to these magnitudes (Weber's law for *subjective* continua).
3. It is assumed that the distributions of all percept magnitudes (and of all their ratios) are identical lognormal distributions.

The above set of conditions corresponds roughly to what Stevens (1959) called case VI. The scale obtained according to case V is a logarithmic function of the case VI scale. The case V scale is also a logarithmic function of a scale obtained by, say, magnitude or ratio estimation (see, for example, Ekman and Sjöberg, 1965). The case VI scale thus agrees, within a power transformation (Stevens, 1975, pp. 83–85), with the corresponding ratio scale, though the exponent cannot be derived from the ordinal judgments on which it is based. (The argument assumes that magnitude and ratio estimation scales are ratio scales.)

Thurstone (1927) is quite explicit that his subjective scale is a defined scale, an "artificial construct" to use his own term. Now we can ask the question: Is the data-equivalence between the two models, case V and case VI, of the trivial kind mentioned in the introduction?

In a way it is, at least within its own scope. Once a choice is made betwen the two conditions 1, both corresponding conditions 2 and 3 have to hold if either model is correct. Whether they do is, however, an empirical matter. Case V may prove empirically wrong; for instance, Ekman's law could hold and the percepts be normally distributed. But then case VI must also be wrong. For the time being, we assume the correctness of case V (and case VI).

I deliberately chose to use the term "interpreted" when describing condition 1. The question is whether that interpretation is made by the researcher, in which case the obtained scale really is a defined one, or by the observer. In the latter case, an ordinal judgment by the observer may be based on a subjective difference or a subjective ratio. (The argument presumes that the ordinal judgment, though apparently simpler than a difference or ratio judgment, is indeed preceded by one of the seemingly more complicated judgments.) Of course, the process in the observer is difficult or perhaps even impossible to identify.

Let us know expand the scope of our scaling problem and consider data gathered by other methods. The conclusion that case VI and ratio scales belong to a power group and thus possibly are the same scale – which does not obtain for the case V scale – favors the case VI scale. One might raise the objection that Fechner's just noticeable difference (JND) scale (a scale obtained by summing JNDs) likewise is a logarithmic function of the corresponding ratio scale (because it is a logarithmic function of the physical scale, and the ratio scale, obtained by, for example, magnitude estimation, is a power function of the physical scale; see, for example, Stevens, 1975). However, the case V and the JND scales are closely related, both being based on measures of variability and both assuming that this measure on the subjective side is constant. Fechner assumed that the subjective increment corresponding to a JND on the physical continuum is constant over the whole stimulus range, and Thurstone assumed that all standard deviations were equal (condition 2 for case V). Thus, the agreement between JND and case V scales is to be expected, whereas the agreement between case VI and magnitude scales (within a power transformation) is not quite as predictable.

The scope of our scaling problem can be expanded in another way that makes possible a choice between our data-equivalent models, a way that may also be applied to other problems. We may investigate whether one but not the other model can be incorporated into a more comprehensive theory. Such a theory for the case at hand is the prediction of choices between compound objects (Bock & Jones, 1968, pp. 263–275). In experiments investigating this theory observers indicate their preference for pairs of objects. The set of objects consists both of single stimuli (e.g., a pen or a pencil) and objects composed of two stimuli (the pen and the pencil together). Each object, single or "composite", is paired with every other object, though pairs consisting of composite and single stimuli in which the single stimulus is part of the composite are omitted (because A and B are always preferred to A alone). The data obtained are the relative frequencies of one object being preferred to another one. In the theory described by Bock and Jones a zero point can be determined and the combination of two stimuli into one object is defined as addition of scale values. From the results of an experiment Bock and Jones (1968, p. 273) concluded: "... the imposition of an additive scale model and collection of data for both single and composite stimuli, does permit a distinction between goodness of fit for case V and case VI. Results of the present analysis argue strongly for case VI ..."

Thus we see that the inclusion of one or the other model in a more comprehensive theory may destroy the models' data-equivalence, thereby making a choice between them possible by comparing their fit. For the case at hand, I am inclined to favor case VI above case V, both because the scale obtained from case VI agrees with a ratio (e.g., magnitude) scale within a power transformation and because of the better fit for choices of compound objects. The data-equivalence turned out not to be trivial and we may tentatively conclude that the observers' ordinal judgments are based on a subjective ratio rather than on a subjective difference (see also Eisler, 1978).

Hellström's experiments (this volume) in which observers are required to use three rather than two categories in pair comparison (the third category being "equal" or "don't know") may open another way of distinguishing between case V and case VI, or of proving both wrong.

The remaining examples are drawn from theories of similarity and multidimensional scaling.

Lack of Uniqueness in City-Block Space

The example now to be presented resembles the one discussed earlier (pp. 12–15) in that the data-equivalence does not obtain for two or more models but for parameter values in a model, because the parameters cannot be determined uniquely.

For an n-dimensional city-block metric (for the present purpose, it suffices to say that a metric defines the distance between two points in a space), the distance d_{ij} between percepts i and j is defined as

$$d_{ij} = \sum_{k=1}^{n} |x_{ik} - x_{jk}|,$$

where x_{ik} is the coordinate of percept i in dimension k and the summation is over all n dimensions. The metric is called city-block because it defines distance (in two dimensions) as the walking distance between two spots in a city with sets of parallel streets at right angles to each other. Given all pairwise distances, the necessary conditions for obtaining a unique solution for the coordinates of all percepts in all dimensions were derived in Eisler (1973). To put it simply, no percept is permitted to have the lowest coordinate of all percepts in more than one dimension, nor the highest coordinate in more than one dimension. Figure 4, taken from Eisler (1975b), illustrates this. The position of point A cannot be determined uniquely; it can move along the straight line A–A′ without any change in the city-block distances to the other three points B, C, and D. The distances are always 4, 3, and 5 units, respectively. This state of affairs is a consequence of point A having both the lowest x- and the lowest y-coordinate.

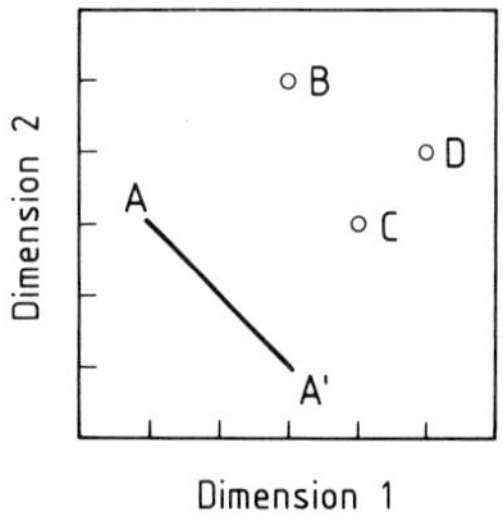

Fig. 4. Illustration of lack, of uniqueness of the coordinates of point A. For details see text. (From Eisler, 1975b)

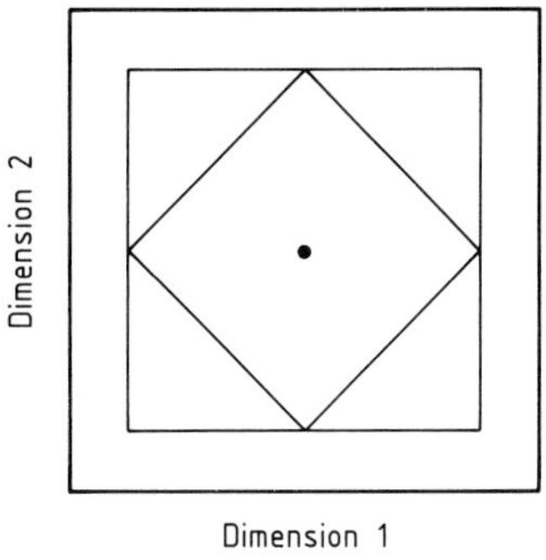

Fig. 5. Data-equivalent isosimilarity contours for the city-block and the dominance metrics. The standard is indicated by a *filled circle*

The situation is, however, easy to remedy. In order to obtain unique solutions for the coordinates, "... the researcher should suppress his urge for symmetry so that no point (stimulus) is, say, smallest in more than one dimension" (Eisler, 1973, p. 216).

Data-Equivalence Between City-Block and Dominance Metrics in Two-Dimensional Space

We are already familiar with the city-block metric. In the dominance (or supremum) metric the distance between two points is defined as the largest distance in any dimension. In the city-block metric with two dimensions, when the difference between the x-coordinates of the two points is 3, and the difference between the y-coordinates 5, the distance becomes 8. In the dominance metric, it becomes 5.

Metrics can be demonstrated by their corresponding isosimilarity contours (Shepard, 1964). An isosimilarity contour is the locus of all points having a given distance from a standard point. (Similarity is assumed to be a one-to-one monotonically decreasing function of distance in the subjective space, so that a given distance value corresponds uniquely to a similarity value.) For the dominance metric the isosimilarity contours are squares, for the city-block metric they are diamonds (see Fig. 5). These two metrics are identical for two dimensions, except for a 45° rotation (Shepard, 1974); that is, given all distances they both fit equally well: data-equivalence. The only way I can see of avoiding an arbitrary choice between these metrics is to complicate the stimuli so that at least three dimensions are obtained. However, the problem often solves itself: even if the subjective space is devised to be two-dimensional, the observers may disagree with the researcher as to the dimensionality and see more than the expected two dimensions (see Eisler & Roskam, 1977b).

Data-Equivalence Between Vector and Set Theoretical Models of Similarity

In the previous section it was mentioned that similarity is assumed to be a monotonically decreasing function of distance in the subjective space. (In certain connections this function may be specified.) The models we have been dealing with belong to a set of *distance models,* to be distinguished from *content models* (Ekman & Sjöberg, 1965). According to Ekman (1961), in content models similarity of a pair of percepts may be defined as a ratio, namely, the common subjective magnitude (taken from both percepts) over the total subjective magnitude. This definition invites a set theoretical formulation:

$$s_{ij} = \frac{2\,\mathrm{m}\,(i \cap j)}{\mathrm{m}\,(i) + \mathrm{m}\,(j)}$$

where s_{ij} is the similarity between percepts *i* and *j,* and m (*i*) the measure of set *i,* regarding percepts as sets. Thus, similarity is, according to the set theoretical formulation of the content model, twice the measure of the intersection of the two percepts over the sum of their measures.

Another way of defining similarity in content models is the vector model, originally devised by Ekman (1954) for purely qualitative variation (equal vector lengths for all percepts) and by Eisler and Ekman (1959) for purely quantitative variation, that is, unidimensional percepts (the angle between vectors of all pairs of percepts is 0°). The equation for multidimensional similarity with unequal vector lengths is

$$s_{ij} = \frac{2\,h_i}{h_i + h_j} \cos v_{ij}, \quad h_i \leq h_j .$$

Here, percepts are seen as vectors in multidimensional space rather than as points, as in the distance models. The vector length of percept *i* is h_i, and the angle between percepts *i* and *j* is v_{ij}. The two equations for similarity according to the content model can be made equivalent by equating set measures with vector lengths, $m\,(i) = h_i$, and the intersection $m\,(i \cap j)$ with $\min\,(h_i, h_j) \cos v_{ij}$; $\cos v_{ij}$ is interpreted as the fraction of the smaller percept that it has in common with the larger (Eisler, 1980; Eisler & Roskam, 1977a, b).

Again, we have two data-equivalent models. In this case, there is no reason to choose one. The set theoretical model is appealing because it seems natural to describe similarity in terms of what is common to two percepts in relation to their totality. The drawback of this description is that it is difficult to visualize the relations between a great number of percepts, any two of which have some overlap. The vector model, on the other hand, allows the familiar description of the percepts as vectors in multidimensional space.

Data-Equivalent Representations of the Content Model for Similarity

In the experiment presented by Eisler and Roskam (1977a, b), each stimulus entering into a pair, the similarity of which was to be judged, consisted of two luminous points forming a "pattern" so that one point could always be seen as located in the origin and the other somewhere in the first quadrant. One of the problems to be dealt with was whether observers attend to the *x*- and *y*-coordinates or to the length and angle of the percepts – you could say,

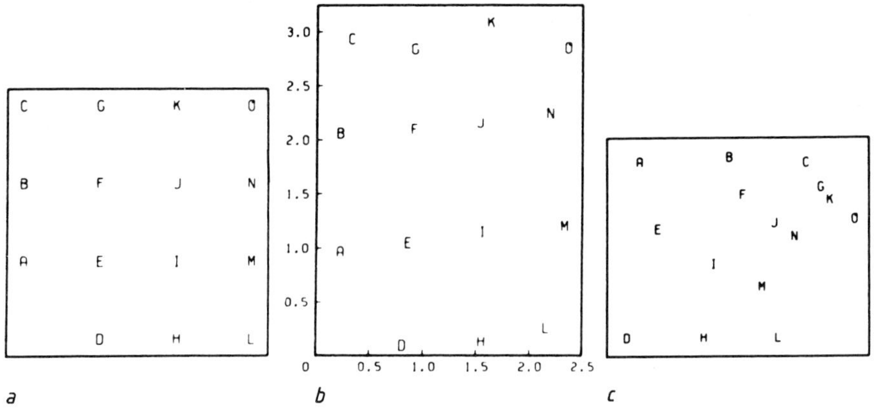

Fig. 6a–c. The physical arrangement of the stimuli (**a**) and the results of multidimensional analyses according to the content model (**b**) and the city-block model (**c**). Stimuli (and percepts) are denoted *A–O*. (Adapted from Eisler & Roskam, 1977b. Copyright 1977 by North-Holland Publishing Co. Reprinted by permission)

whether observers see the two-point patterns in terms of Cartesian or polar coordinates.

Unfortunately, the experiment did not solve this problem. Both the content and the city-block models were applied in multidimensional scaling of the data. The content model yielded a subjective space in accordance with the Cartesian representation, and the city-block model in accordance with the polar one. To be more specific: the representation according to the content model looked like the physical arrangement (with small distortions), that is, points in the first quadrant, with the horizontal as x- and the vertical as y-axis. The city-block representation, on the other hand, did not look like the physical point arrangement at all: the first dimension came out as length (subjective distance between the two points), the second as angle (subjective angle between the x-axis and the line connecting the two points). Thus the polar coordinates of the content model representation were represented as Cartesian in the city-block model (length on the x- and angle on the y-axis). Figure 6 makes this clear. Figure 6a shows the physical arrangement of the point pairs, denoted A–O. The letters are placed at the positions of the variable points, with the fixed point (at the origin) omitted. Figure 6b shows the space obtained by applying the content model, and Fig. 6c the result of multidimensional scaling according to the city-block model. If we look at the percepts in Fig. 6b in terms of polar coordinates, we see that percept O, for example, has the largest vector length and the angle is in the middle of the range. This agrees with the x- and y-coordinates for this percept in Fig. 6c. Percepts A and D have the shortest vector lengths and the extreme angles with the x-axis, 0° and 90°. Again, this is reflected in Fig. 6c. The fit was good for both the distance and the content model and provided no basis for a choice. (These two models are *not* data-equivalent.) Can the outcome of the content-model scaling be made similar to that of the city-block scaling? This is simply a question of how we wish to represent our data. We can compute the vector lengths and angles from the content-model plot, and replot them as Cartesian coordinates (see Eisler, 1980). We may tentatively equate the vector length, that is, the x-coordinate in the new plot, with quantity (e.g., subjective intensity or impressiveness), and the angle, the new y-coordinate, with quality.

In a content-model representation like that in Fig. 6b, the origin, and thereby the vector lengths, is fixed, as opposed to the position of the axes, as component analysis is used for the scaling. The axes can be rotated arbitrarily relative to the set of vectors, so that the angle between an axis and a percept vector is arbitrary, while the angle between any pair of vectors is fixed. This corresponds to the similarity equation for the content model, which is built on the vector lengths h_i and h_j and the angle between the vectors of this pair of percepts, v_{ij}.

Assume that as stimuli, we have tones varying in amplitude and frequency. Loudness should come out as vector length – or as the x-coordinate in the new representation. The range of pitches should span a 90° angle; if the tone frequencies varied between, say, 200 and 5000 Hz, we should obtain a 200-Hz and a 5000-Hz "factor". In the new plot, the y-coordinate would vary between the pitches corresponding to 200 and 5000 Hz, a representation that seems to make more sense, though data-equivalent with the one obtained directly by means of component analysis.

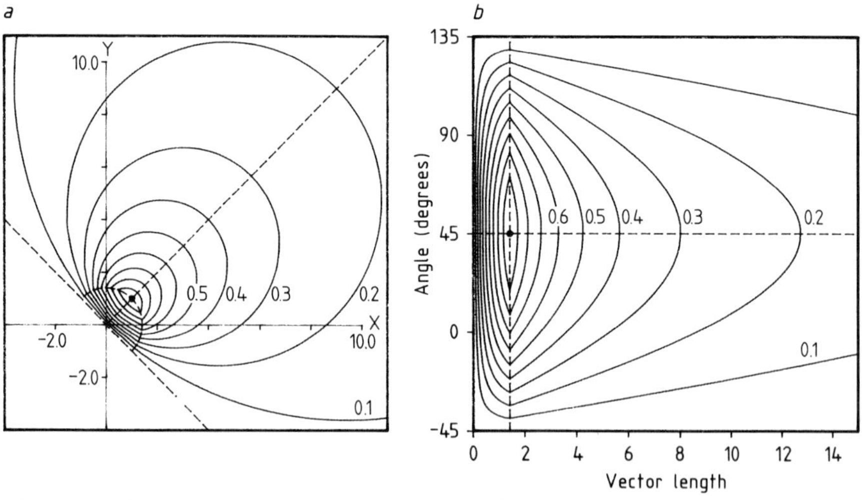

Fig. 7a, b. Theoretical isosimilarity contours according to the content model as obtained directly from **a** component analysis and **b** after replotting vector lengths on the x- and angle on the y-axis. The standard is indicated by a *filled circle*. The *numbers* are similarity values on a scale between 0 and 1

Figure 7 shows isosimilarity contours for the content model, Fig. 7a as it comes out from multidimensional analyses (see also Gregson, 1975, p. 123), Fig. 7b after replotting of polar coordinates as Cartesian.

These two representations are, of course, data-equivalent. Whether the data-equivalence is of the trivial kind mentioned in the introduction may be a moot question.

Conclusion

In the above discussion we have seen different kinds of data-equivalence. There were illustrations of simple cases where it merely proved impossible to determine model parameters uniquely, and more elaborate ones for which the models themselves were data-equivalent. If a choice between data-equivalent models is desired, it sometimes helps to extend the data set or embed the models in a more comprehensive theory. Other criteria such as expedience can also be invoked. Finally, different representations might bring out different aspects of one's data, perhaps making a choice undesirable.

References

Allan, L.G. (1977). The time-order error in judgments of duration. *Canadian Journal of Psychology, 31*, 24–31.

Bock, R.D. & Jones, L.V. (1968). *The measurement and prediction of judgment and choice.* San Francisco: Holden-Day.

Eisler, H. (1965a). The connection between magnitude and discrimination scales and direct and indirect scaling methods. *Psychometrika, 30,* 271–289.

Eisler, H. (1965b). On psychophysics in general and the general psychophysical differential equation in particular. *Scandinavian Journal of Psychology, 6,* 85–102.

Eisler, H. (1973). The algebraic and statistical tractability of the city block metric. *British Journal of Mathematical and Statistical Psychology, 26,* 212–218.

Eisler, H. (1974). The derivation of Stevens' psychophysical power law. In H.R. Moskowitz, B. Scharf & J.C. Stevens (Eds.), *Sensation and Measurement* (pp. 61–64). Dordrecht, Holland: Reidel.

Eisler, H. (1975a). Subjective duration and psychophysics. *Psychological Review, 82,* 429–450.

Eisler, H. (1975b). The uniqueness problem in metric and non-metric MDS models and hazards in the use of ordinal proximity scales in MDS scaling. *Göteborg Psychological Reports, 5* (29), 8–9.

Eisler, H. (1978). On the ability to estimate differences: A note on Birnbaum's subtractive model. *Perception and Psychophysics, 24,* 185–189.

Eisler, H. (1980). Equivalence of psychological models: Time-order errors, similarities, subjective spaces. In E.D. Lantermann & H. Feger (Eds.), *Similarity and choice* (pp. 88– 99). Bern: Huber.

Eisler, H. (1981a). Applicability of the parallel-clock model to duration discrimination. *Perception and Psychophysics, 29,* 225–233.

Eisler, H. (1981b). The parallel-clock model: Replies to critics and criticisms. *Perception and Psychophysics, 29,* 516–520.

Eisler, H. & Ekman, G. (1959). A mechanism of subjective similarity. *Acta Psychologica* (Amsterdam), *16,* 1–10.

Eisler, H. & Roskam, E. E. (1977 a). Multidimensional similarity: An experimental and theoretical comparison of vector, distance, and set theoretical models. I. Models and internal consistency of data. *Acta Psychologica* (Amsterdam), *41,* 1–46.

Eisler, H. & Roskam, E. E. (1977 b). Multidimensional similarity: An experimental and theoretical comparison of vector, distance, and set theoretical models. II. Multidimensional analyses: The subjective space. *Acta Psychologica* (Amsterdam), *41,* 335–363.

Ekman, G. (1954). Dimensions of color vision. *Journal of Psychology, 38,* 467–474.

Ekman, G. (1959). Weber's law and related functions. *Journal of Psychology, 47,* 343–352.

Ekman, G. (1961). Some aspects of psychophysical research. In W.A. Rosenblith (Ed.), *Sensory communication* (pp. 35–47). New York: Wiley.

Ekman, G. & Sjöberg, L. (1965). Scaling. *Annual Review of Psychology, 16,* 451–474.

Gregson, R.A.M. (1975). *Psychometrics of similarity.* New York: Academic.

Shepard, R.N. (1964). Attention and the metric structure of the stimulus space. *Journal of Mathematical Psychology, 1,* 54–87.

Shepard, R.N. (1974). Representation of structure in similarity data: Problems and prospects. *Psychometrika, 39,* 373–421.

Stevens, S.S. (1959). Sic transit gloria varietatis? (Review of L. L. Thurstone's "The measurement of values".) *Contemporary Psychology, 4,* 388–389.

Stevens, S.S. (1975). *Psychophysics.* New York: Wiley.

Suppes, P. & Zinnes, J. L. (1963). Basic measurement theory. In R. D. Luce, R. R. Bush, & E. Galanter (Eds.), *Handbook of mathematical psychology* (Vol. 1, pp. 1–76). New York: Wiley.

Thurstone, L. L. (1927). A law of comparative judgment. *Psychological Review, 34,* 273–286.

What Happens When We Compare Two Stimuli?*

Åke Hellström

Abstract

Comparing two stimuli involves forming, for each stimulus, a weighted compound of the sensation magnitudes of the stimulus and of the current adaptation level, and taking the difference between the two compounds. Although this causes systematic errors, it can greatly increase the detectability of a small change in the difference between the stimuli. The optimal weighting reflects the amount of information available about the stimuli. Computerized neuropsychological tests built on this notion have proved clinically useful. The model also throws light on questions related to psychophysical laws and subjective variability.

The Time-Order Error

Introduction

Comparing two stimuli is a much more complex task than one might think. Take a situation where the stimuli are presented with a short time interval between them, and the subject is to tell, for example, which is the greater. Not only will the judgments fluctuate, but even if the stimuli are physically equal, they may be systematically judged as different, so that to make them seem equal on average, one of them has to be changed. This phenomenon was discovered by the founder of psychophysics, Gustav Fechner (1860), who named it *Zeitfehler* (*time-order error,* TOE) and defined it as negative when the first stimulus is underestimated relative to the second, and as positive in the opposite case.

TOE – Perception or Response?

It was clear from the start that the TOE is not a simple phenomenon: Fechner likened it to the giant Proteus with his many disguises, and later researchers have described it using terms like "elusive" and "idiosyncratic." It has been

---

* This research was supported by the Swedish Council for Research in the Humanities and Social Sciences and by the Åke Wiberg Foundation.

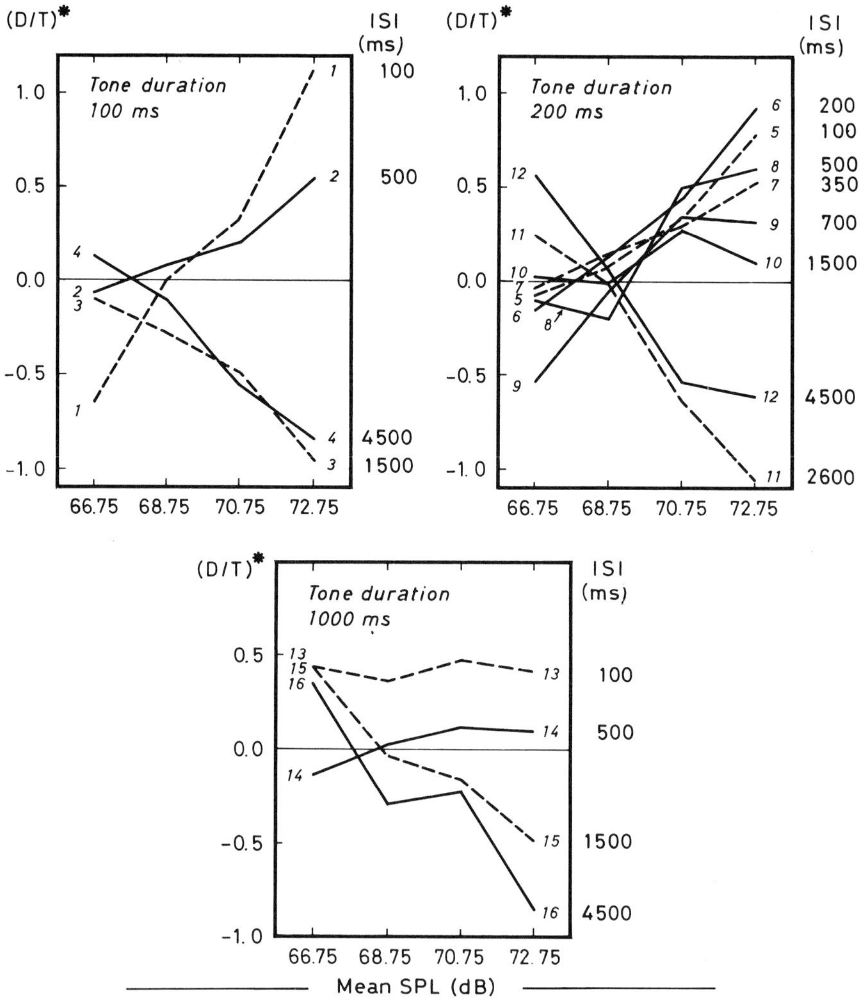

Fig. 1. Mean estimated loudness difference D in units of subjective *equal* category half-width T for pairs with SPL difference ± 0.5 dB versus mean SPL for each condition (group). Group number is indicated at the *ends* of each curve, ISI to the *right* of each panel. Some of the lines are dashed for clarity. (From "Time errors and differential sensation weighting" by A. Hellström, 1979, *Journal of Experimental Psychology: Human Perception and Performance, 5,* 460–477. Copyright 1979 by the American Psychological Association. Reprinted by permission)

debated whether the TOE is a perceptual phenomenon, as held by Fechner and his followers, or a product of response bias, as has later become a popular notion. Restle (1961) argued on set-theoretical grounds that judgments like *louder* have an inherent advantage over judgments like *softer,* which (as it is usually the second stimulus that is the "comparison" and the first the "standard") would explain why negative TOEs are more common than positive. In two studies (Hellström, 1977, 1978) the judgment mode proved rather unimportant: when the first stimulus was judged in relation to the second, the

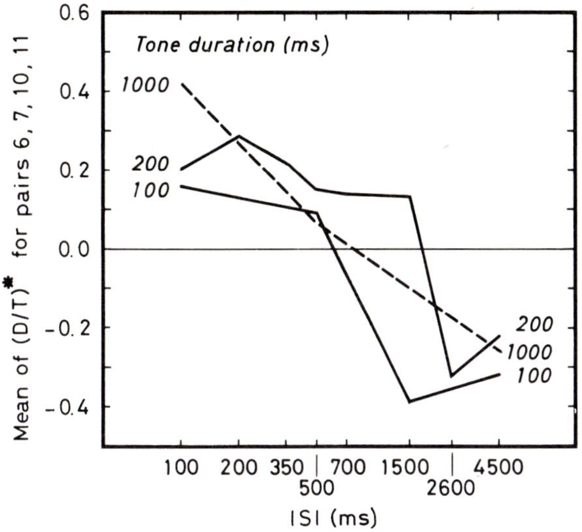

Fig. 2. Mean estimated loudness difference D in units of subjective *equal* category half-width T for pairs with SPL difference ± 0.5 dB versus ISI (tone duration indicated at the *ends* of each curve). (From "Time errors and differential sensation weighting" by A. Hellström, 1979, *Journal of Experimental Psychology: Human Perception and Performance, 5,* 460–477. Copyright 1979 by the American Psychological Association. Reprinted by permission)

TOE pattern was the same as with the usual procedure. In agreement with classical results, TOEs were negative for higher, and positive for lower levels of stimulus magnitude. It seems that response bias and similar explanations can be ruled out, the TOE belonging in the perceptual/cognitive realm (for a discussion see Hellström, 1985).

Loudness Comparison: A Factorial Experiment

What is it that happens, then, when two successive stimuli are compared? To find out, one must dispose of the traditional designs that yield "constant errors," which are by no means constant, and turn instead to factorial designs. In Hellström (1979), combining four values of mean sound-pressure level (SPL) and four levels of intrapair SPL difference yielded 16 tone pairs. Also, there were 16 combinations of interstimulus interval (ISI) and tone duration, with one 12-subject group for each condition (judging each pair six times). The question was "Which tone was louder?"; judgments of *equal* were allowed. Applying a Thurstonian method to the group data, the mean subjective difference, D, in each pair was scaled in units of the subjective half-width, T, of the *equal* category. As Fig. 1 shows, there were strong stimulus-level effects on the TOE (defined as D/T for two physically equal stimuli), these effects depending on the ISI and tone duration. For short durations and ISIs, the usual effect was reversed. Figure 2 demonstrates that (as in the classical studies) the mean TOE was positive for short, and negative for longer ISIs.

Sensation-Weighting Models for the Subjective Difference

Model 1, fitted by multiple regression to the data from each condition, explained 96.5% of the total variance:

$$D/T = W_1 \Psi_1 - W_2 \Psi_2 + C \,, \tag{1}$$

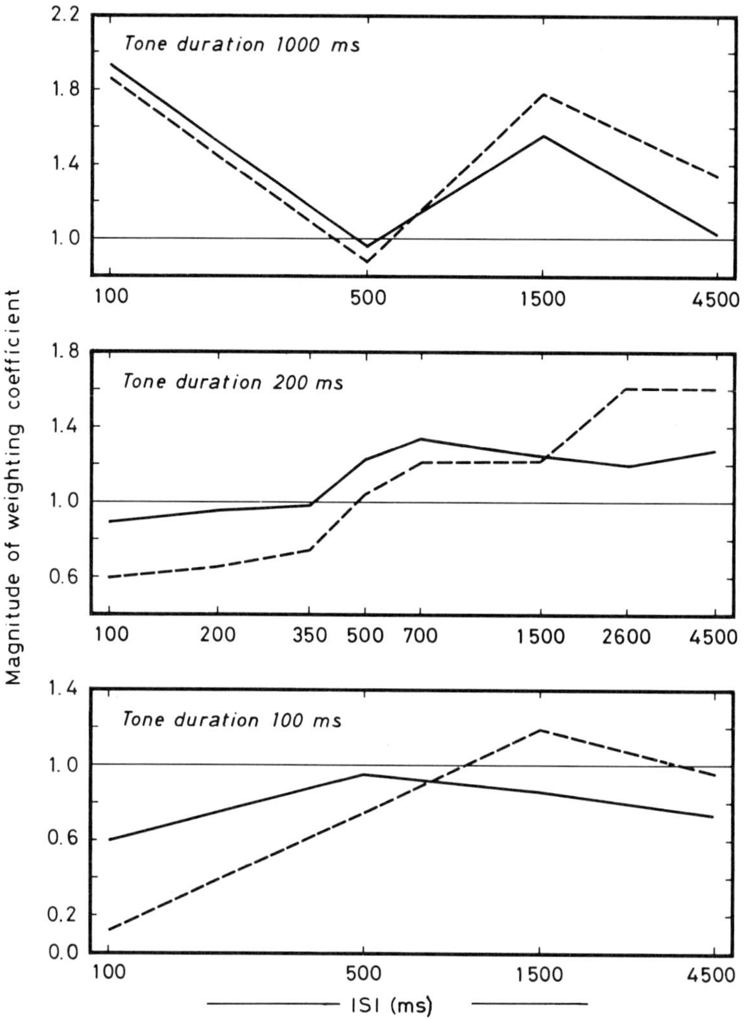

Fig. 3. Stimulus weights s_1 (*continuous lines*) and s_2 (*dashed lines*) in Model 3 versus ISI for each tone duration. (From "Time errors and differential sensation weighting" by A. Hellström, 1979, *Journal of Experimental Psychology: Human Perception and Performance, 5,* 460–477. Copyright 1979 by the American Psychological Association. Reprinted by permission)

where Ψ_1 and Ψ_2 are the sensation magnitudes of the compared stimuli (for their relation to the physical magnitudes, a power function was fitted). W_1 and W_2 are the corresponding regression weights and C is the intercept. A strong linear dependence over the 16 conditions of C on W_1 and W_2 indicated that higher-order models could be devised. Model 3 (a generalized version of the model of Michels and Helson, 1954) explained 95% of the total variance:

$$D/T = k\{[s_1\,\Psi_1 + (1-s_1)\,\Psi_{a1}] - [s_2\,\Psi_2 + (1-s_2)\,\Psi_{a2}]\}\,, \tag{2}$$

where k is a scale factor, s_1 and s_2 are the stimulus weights, and Ψ_{a1} and Ψ_{a2} are the sensation magnitudes that correspond to the adaptation levels (ALs) pertinent to the two stimuli. (The two ALs were slightly different and were located below the center of the stimulus set.) Figure 3 shows the fitted s values, plotted against the ISI for each tone duration.

The Origin of the TOE

Some theorists have tried to explain the (negative) TOE in terms of memory traces becoming weaker with time, so that the first stimulus seems softer, lighter, smaller, etc., than the second. According to Köhler (1923), each stimulus leaves a physiological trace that first waxes, then wanes. This was supposed to explain the shift from positive to negative TOEs with increasing ISI (see Fig. 3). However, this explanation seems too simplistic: the clarity with which we remember a tone need not affect how loud we remember it to have been.

Could our quantitative modeling help us understand how the TOE arises? Setting, in Model 3 (Eq. 2), $\Psi_1 = \Psi_2 = \Psi$, we obtain a measure of the TOE in subjective units. Also setting $\Psi_{a1} = \Psi_{a2} = \Psi_a$ yields

$$\mathrm{TOE} = k\,(s_1 - s_2)\,(\Psi - \Psi_a)\,. \qquad (3)$$

We see from this that the source of the stimulus-level effect on the TOE, and the main factor behind the TOE in general, is the nonequality of s_1 and s_2, in combination with the difference between the stimuli and their ALs.

Such results as those shown in Fig. 3, where s_1 decreases towards longer ISIs, now suggest that the changing-trace theory contains a kernel of truth, in that the stimulus weights (s values), and thus the TOE, reflect the integration of information at the moment of comparison: just as in other cognitive tasks, specific information and generic information are weighted together. Here, the information sources are the stimuli and the ALs, rspectively. The less specific (stimulus) information available, the more generic (AL) information is weighted in. This simple principle helps us to understand the TOE and the comparison process in general (see Hellström, 1985). However, there is more to be said on the subject.

Optimizing Sensitivity to Change

Introduction

As is well-known, our perceptual mechanisms are specialized for the detection of *changes,* not only in stimulus levels, but also in stimulus patterns, gradients, and relations. These convey important information; for instance, a gradient over space or time may indicate increasing distance or gradually decreasing illumination. The presence and movement of objects, and the occurrence of events, are indicated by changes in gradients, or, more generally, in stimulus relations. The Mach band phenomenon illustrates how gradient changes are perceptually highlighted, even at the expense of veridicality.

Maximizing Signal-to-Noise Ratio for Changes in Stimulus Differences

In a comparison of two stimuli, one important aim of the perceptual system must be to maximize the detectability of nonrandom changes in the physical difference between the stimuli. This requires that the change in their *experienced* difference d become as large as possible, relative to the noise. In other words, perception must strive to maximize the signal-to-noise ratio (S/N) of d, that is, $\Delta d/\sigma_d$, where Δd is the nonrandom change, or signal magnitude, of d, and σ_d its standard deviation, or noise magnitude. (Gulliksen, 1968, termed σ_d the *comparatal dispersion.*) The reason for the occurrence of Δd must be that Ψ_1 and/or Ψ_2 has changed. We thus get

$$S/N = \frac{u\,\Delta\Psi_1\,(\partial d/\partial\Psi_1) + (1-u)\,\Delta\Psi_2\,(\partial d/\partial\Psi_2)}{\sigma_d}, \tag{4}$$

where $\Delta\Psi_1$ and $\Delta\Psi_2$ are the changes in Ψ_1 and Ψ_2, and $\partial d/\partial\Psi_1$ and $\partial d/\partial\Psi_2$ the partial derivatives of d with respect to Ψ_1 and Ψ_2. u reflects the relative attention to the two stimuli. Simplifying by setting u to its neutral value of 0.5, $\Delta\Psi_1$ to $+1$, and $\Delta\Psi_2$ to -1 yields

$$S/N = (s_1 + s_2)/(2\,\sigma_d)\,. \tag{5}$$

Expressing σ_d in terms of Model 3 (Eq. 2), we may calculate the values of s_1 and s_2 that maximize S/N (see Hellström, 1986a). These optimal s values are determined by the σ values and intercorrelations (ϱ) of ($\Psi_1 - \Psi_{a1}$), ($\Psi_2 - \Psi_{a1}$), and ($\Psi_{a1} - \Psi_{a2}$). These in turn are functions of the σ and ϱ values of Ψ_1, Ψ_2, Ψ_{a1}, and Ψ_{a2}.

Table 1 gives the optimization results for a few cases. The gain percentage, $G\%$, is the increase in S/N, as a percentage of the S/N that would result with $s_1 = s_2 = 1$ (i.e., ignoring the ALs). Three series of cases are shown: series 1 and 2 are selected in an attempt to represent realistic conditions in successive comparing: series 3 illustrates what happens when all intercorrelations are zero. For each series, the values of σ and ϱ for Ψ_1, Ψ_{a1}, and Ψ_{a2} are assumed to be fixed, while $\sigma_{\Psi 2}$ varies. The optimized s ratio varies with $\sigma_{\Psi 2}$. The last two columns show the optimized values of $(s_1 - s_2)^2$ and the ensuing σ_d^2 values. In Hellström (1979) their empirical values were strongly correlated, which could be explained by assuming σ_Ψ to be equal for the two stimuli and constant over conditions, and $\varrho_{\Psi 1,\Psi 2}$ to be close to 1. However, as the last two columns of Table 1 indicate, the discrimination optimization model predicts similar results over a wide range of conditions.

Memory loss, stimulus interference, or reduced attention capacity may add noise to, and thus increase the variability of, one or both of the compared sensation magnitudes. This, like the intercorrelations of stimulus and/or AL magnitudes, calls for perceptual sharpening by optimal stimulus weighting. As is demonstrated by the high values of $G\%$, this discrimination optimization principle is powerful in helping the perceiver make more efficient use of incoming information, and thus has a clear survival value. Most likely, therefore, it

Table 1. Optimal s_1, s_2, signal-to-noise ratio (S/N), and S/N gain percentage ($G\%$) according to the discrimination optimization model

$\sigma_{\psi 2}$	Optimal s_1	Optimal s_2	s_1/s_2	S/N	$G\%$	$(s_1-s_2)^2$	σ_d^2
Series 1							
0.50	1.25	1.56	0.80	3.49	134	0.095	0.162
1.00	1.67	1.67	1.00	5.00	216	0.000	0.111
1.10	1.76	1.69	1.04	4.67	213	0.005	0.136
2.00	2.70	1.89	1.43	2.16	189	0.660	1.135
3.00	4.05	2.14	1.89	1.41	221	3.627	4.840
Series 2							
0.50	0.75	0.93	0.81	2.88	70	0.031	0.085
1.00	0.79	0.79	1.00	5.00	124	0.000	0.025
1.10	0.80	0.77	1.04	4.61	121	0.001	0.029
2.00	0.86	0.60	1.44	1.82	115	0.069	0.159
3.00	0.91	0.47	1.94	1.16	149	0.194	0.355
Series 3							
0.50	0.87	1.13	0.77	0.93	4	0.068	1.152
1.00	1.00	1.00	1.00	0.71	0	0.000	2.000
1.10	1.03	0.97	1.06	0.67	0	0.003	2.803
2.00	1.32	0.68	1.92	0.50	11	0.399	4.053
3.00	1.55	0.45	3.46	0.42	34	1.218	5.586

Common assumptions: $\sigma_{\psi 1}=1$; $\sigma_{\psi a1}=\sigma_{\psi a2}=1.5$.
Further assumptions:

Series	1	2	3
$\varrho_{\psi 1,\psi 2}$	0.8	0.9	0.0
$\varrho_{\psi 1,\psi a1}$	−0.4	−0.5	0.0
$\varrho_{\psi 2,\psi a2}$	−0.4	−0.5	0.0
$\varrho_{\psi 1,\psi a2}$	−0.7	−0.3	0.0
$\varrho_{\psi 2,\psi a1}$	−0.7	−0.3	0.0
$\varrho_{\psi a1,\psi a2}$	0.5	0.5	0.0

was built in by evolution. The normally rather slight systematic errors that result from it should not be too high a price to pay, considering the advantages.

A Parallel: Brunswik's Lens Model

The discrimination optimization model is related to Brunswik's lens-model account of perceptual judgment (Postman & Tolman, 1959; cf. Hake, Rodwan & Weintraub, 1966). According to this view, the aim of perception is to maximize the correlation between actual and judged properties of distal stimuli. To achieve this, each proximal cue should be weighted in accordance with its correlation with the stimulus, its "ecological validity" (Hursch, Hammond, & Hursch, 1964). One factor behind this validity of the cue is its freedom from random errors. Regarding each cue as a sample of microcues, the size of this sample (the amount of information it carries) will affect the standard error of the mean cue magnitude. The correlation of the judgment with an information-rich cue can be raised by increasing the relative weight of this cue in determining the judgment. In the comparison task, discrimination optimization employs the same principle for each of the two stimuli, regarding it as a cue to the intrapair difference.

Neuropsychological Applications of the Sensation-Weighting Models

Background

In many brain disorders, the intake of stimulus information is disturbed. The discrimination optimization model implies that this should lead to altered stimulus weighting in comparing, which in turn suggests that TOE tests could be diagnostically useful. In fact, differences in TOE have been found between brain-damaged patients and normal subjects (e.g., Birch, Belmont & Karp, 1965; Van Allen, Benton & Gordon, 1966; Needham & Black, 1970). However, these differences are hard to interpret as they run in different directions. This is hardly surprising: the TOE is not very informative in itself; one must get down to the underlying processes.

Solvent-Related Brain Dysfunction

Hellström, Åslund, and Almkvist (1985) studied workers who had been exposed to organic solvents for at least 8 years and who were being investigated for suspected solvent-related cerebral dysfunction. This "painter's syndrome" is partly characterized by impaired visuospatial memory and lowered perceptual-motor speed. Two "TOE tests" were used, one with line lengths and another with tone durations. The stimuli were presented by a microcomputer, the ISI and both stimulus magnitudes being varied. Each test took about 10–20 min. Because the data were insufficient for model 3, model 1 was fitted. Its stimulus weights W_1 and W_2 are proportional to s_1 and s_2. Assuming linear psychophysical functions, the physical stimulus values were entered directly. The scaling of the perceived difference was also simplified.

Figure 4 shows the main line-length test results, in terms of W values, from one part of the study. Patients diagnosed (using conventional tests) as having a solvent-related brain dysfunction made fewer correct judgments and had lower multiple correlations between stimuli and judgments as well as lower values of W_1 and W_2, than those diagnosed as not having this dysfunction. For both stimulus types and in both patient groups, W_1 changed much more with the ISI than did W_2. For line length, both W_1 and W_2 decreased with increasing ISI. In the dysfunctional patients this trend was more pronounced for W_1 than for W_2, which suggests that forgetting the first line was the dominant process. This effect was absent in the nondysfunctional patients ($p < .05$ for the inter-

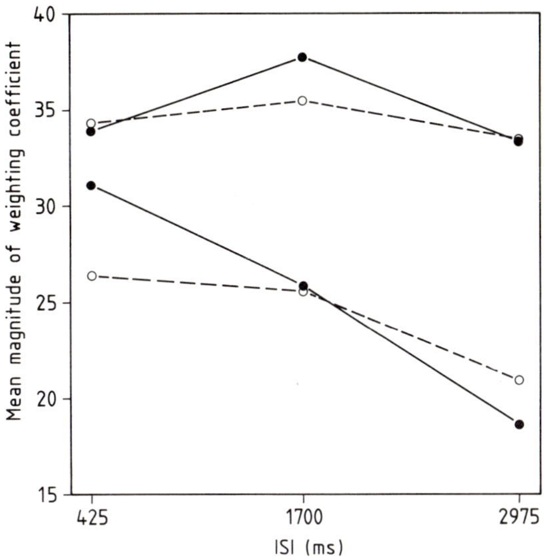

Fig. 4. Line length test: stimulus weights W_1 (*continuous lines*) and W_2 (*dashed lines*) in model 1 versus ISI for patients diagnosed as not having (*upper lines*) and as having solvent-related brain dysfunction (*lower lines*). (Unpublished data from Hellström, Åslund & Almkvist 1985)

group difference). Thus, solvent-related brain dysfunction was once more shown to be associated with accelerated forgetting of visuospatial information.

Senile Dementia

In another study (Hellström, Forssell, & Fernaeaus, 1986) patients with senile dementia of the Alzheimer type (SDAT) in its milder initial stage (mainly characterized by memory problems) were compared with healthy persons of the same age (about 80 years). Here, the tone duration test gave the clearest

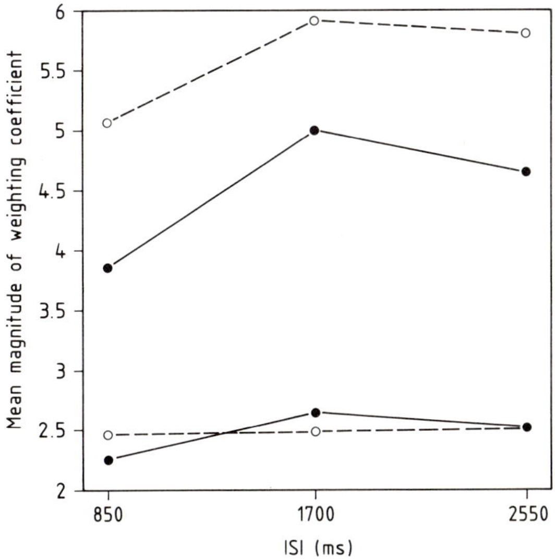

Fig. 5. Tone duration test: stimulus weights W_1 (*continuous lines*) and W_2 (*dashed lines*) in model 1 versus ISI for healthy controls (*upper lines*) and SDAT patients (*lower lines*). (Unpublished data from Hellström, Forssell, & Fernaeus, 1989)

indications. Besides the percentage of correct judgments and the stimulus-judgment correlation, the difference between W_1 and W_2 was associated with the presence of SDAT. For the controls, the second duration was more heavily weighted than the first ($W_2 > W_1$). This was also true for both groups in the above-mentioned solvent study and for the subjects (psychology students) of Hellström (1977). However, for the SDAT patients, the two tones were weighted about equally ($W_2 \approx W_1$; see Fig. 5). The intergroup difference in this weighting balance was highly significant ($p < .002$). A likely explanation is that the level of attention to the second duration was lowered in the SDAT patients, due to reduced attention capacity. Assuming the two ALs to be equal, their common value could be estimated from W_1, W_2, and C. The mean AL was significantly lower for the SDAT patients than for the controls ($p = .029$), which suggests that the SDAT patients failed to adapt adequately to the tone durations employed.

Clinical Applicability of TOE Tests

The achievement measures (per cent correct and multiple correlations) were very sensitive to both kinds of dysfunction studied and could therefore not have been used for differential diagnosis. On the other hand, the functional measures (based on the stimulus weights), although less sensitive, yielded specific profiles for the two illnesses. The two TOE tests therefore constitute a rudimentary battery for differential diagnosis of cerebral dysfunction. New TOE tests (e.g., with visual durations as stimuli) and improved testing procedures are being tried out in order to combine sensitivity and specificity.

Cerebral Lateralization Mirrored by Comparing Simultaneous Stimuli

In the above, the comparing process was discussed in terms of successive stimuli. However, Fechner (1860) also mentioned a *Raumfehler,* or *space-order error* (SOE), in comparing two stimuli that are presented simultaneously, one to the left and one to the right. Very little research has been done on the SOE. Most likely, it has the same explanation as the TOE: the two stimuli are given different weights because of the need to sharpen the sensitivity to stimulus change in a situation with unequal amounts of information about the two stimuli. But why would this happen when the obvious time factor is absent?

When the subject is to tell which of two visual objects, one to the left and the other to the right, is the larger, primary sensory processing for each object takes place in the contralateral visual cortex. Higher processing of visuospatial information occurs mainly in the right hemisphere. In the comparison process, it should therefore be easier to access information on the left-hand object than on the right-hand one. According to the discrimination optimization model, the weight (s value) of the left-hand object should then be the greater. I have collected some preliminary data on the comparing of computer-presented horizontal lines. In at least 80% of subjects the weight for the left line is greater (typically about 4% but sometimes considerably more) than that for the right. Additional studies are under way. One interesting prospect is diagnosing lateralized brain damage, which may cause changes in the left–right balance of stimulus weights.

The Comparatal Dispersion

Equal Judgments and Ekman's Law

If we assume that Stevens' (1957) power law ($\Psi = \alpha\,\Phi^{\beta}$) relates physical stimulus magnitudes (Φ) and the sensation magnitudes (Ψ) they evoke, then we have $d\Psi/d\Phi = \beta\,\Psi/\Phi$, or $d\Psi/\Psi = \beta\,d\Phi/\Phi$. If we further assume that Weber's law ($\Delta\Phi = k_1\,\Phi$) relates physical stimulus magnitudes and their "just noticeable" differences, then we have $\Delta\Psi/\Psi = k_1\,\beta$, which is *Ekman's law* (Ekman, 1956, 1959) follows. This law says that the change in sensation corresponding to a just noticeable stimulus change is proportional to the sensation level. As "just noticeable" means that the change in the subjective difference d corresponds to the 75th percentile of its distribution, one consequence is that $\sigma_d = k_2\,\mu_\Psi$. If $\varrho_{\Psi 1,\Psi 2}$ does not change with the sensation level, it will then also be true that $\sigma_\Psi = k_3\,\mu_\Psi$. An extension of this is the so-called case VI (see Eisler, 1965) according to which log Ψ is normally distributed with constant standard deviation. The support for Ekman's law cited has been rather indirect. However, some more direct evidence exists. In Hellström (1977, 1978) I noted that when stimulus pairs on two magnitude levels were presented intermixedly, *equal* judgments were fewer at the higher than at the lower level (even when only pairs with equal amounts of, say, *heavier* and *lighter* judgments were considered). This trend can also be seen in the results of Needham (1935). In a recent, as yet unpublished study of mine, lifted successive weights were compared using the Constant Method with the question, "Which weight is heavier?". Permitted judgments were *first, second,* and *equal* (three categories) or *first-sure, first-believe, first-guess, second-guess, second-believe,* and *second-sure* (six categories). In the Fixed Standards condition, series with standards of 100, 200, and 300 g were run in the same session, one series at a time and in different orders for the subjects. In the Random Standards condition, pairs from all series alternated in random order. With random standards and three judgment categories, the proportion of *equal* judgments decreased with increasing stimulus level for most subjects. The comparatal dispersion σ_d, estimated in units of the subjective half-width T of the *equal* category, increased accordingly. In the six-category condition this was true with respect to, for example, the combined *guess* categories. In contrast, with fixed standards the number of *equal* judgments, and hence σ_d/T, was rather constant over stimulus levels.

The most likely interpretation of these results seems to be that σ_d increases with the stimulus level, as prescribed by Ekman's law, and that when different levels are intermixed, the subjective width ($2\,T$) of the *equal* or *guess* category is not readjusted between levels. This then leads to a decrease in the proportion of *equal* or *guess* judgments with increasing stimulus level.

Equal Judgments and Fechner's Law

MacKay (1963) and Ekman (1964) sugggested that the psychophysical power law, which describes the results of "ratio scaling" experiments, derives from a match between two sensation magnitudes, one evoked by a stimulus and the other by a number, and each obeying Fechner's (1860) law ($\Psi = a \log \Phi + b$). However, assuming the power law for this relation would lead to the same relation between stimuli and judgment. It thus seems that we have two data-equivalent models (Eisler, this volume). However, the above-mentioned *equal* judgments help us to decide between them.

The "internal" Fechner law implies $d\Psi/d\Phi = a/\Phi$ which, together with Weber's law, yields $\Delta\Psi = ak$. Hence, σ_d should be constant over stimulus levels. If T is also constant, then this predicts a roughly *constant* proportion of *equal* judgments, independently of stimulus level. Another attractive possibility could be that T corresponds to a constant sensation ratio $q = (\Psi + T)/\Psi$, which would imply $T = (q-1)\,\Psi$, that is, that the *equal* or *guess* category widens with increasing stimulus level. With constant σ_d, the proportion of *equal* judgments would then *increase*. The observed *decline* of this proportion would require a narrowing down of the *equal* category. Until someone gives a rationale for this possibility, the findings cast severe doubt on Fechner's law as a candidate for the relation between physical and sensation magnitudes, and instead favor the power law.

Applications to the Psychophysics of Scaling

Optimizing Sensitivity to Disequilibrium

As exemplified by the matching model of MacKay and Ekman, it seems fruitful and clarifying to analyze numerical judgment tasks not in terms of numbers evoked by sensations evoked by stimuli, but in terms of the matching of sensation magnitudes. In fact, any psychophysical task can be conceived as the forming, according to task-specific rules, of an equilibrium between sensation magnitudes, some generated externally (by "stimuli") and others internally (by "responses") and all subject to effects of adaptation and weighting. The discrimination optimization concept can then be generalized to a concept of maximizing the detectability of deviations from equilibrium (Hellström, 1986b).

Subjective Differences and Physical Stimulus Ratios

In order to maximize S/N for temporary changes in the stimulus difference, s_1/s_2 should be nearly proportional to $\sigma_{\Psi 2}/\sigma_{\Psi 1}$ (see Table 1, series 3). If the two stimuli lie on the same continuum, and if σ_Ψ is proportional to μ_Ψ (Ekman's law), then for two stimuli far apart on the continuum the optimal s_1/s_2 will also be nearly equal to Ψ_2/Ψ_1. Denoting Ψ_1/Ψ_2 as Q, $(\partial Q\,\partial\Psi_1)/(\partial Q/\partial\Psi_2) = -1/Q$. This means that with $s_1/s_2 = 1/Q$, d will change if and only if Q changes. If the power law holds, then this will happen if and only if the physical stimulus ratio Φ_1/Φ_2 changes (see Yilmaz, 1967). Changes in this ratio carry important information about the environment, and must be sorted out from general changes in, for instance, illumination, which alters the two physical magnitudes proportionally. The power law and the discrimination optimization principle complement each other in achieving this.

Teghtsoonian's Laws

R. Teghtsoonian (1971, 1973) has formulated two important generalizations, which may well be called laws (see Hellström, 1986 b). According to *Teghtsoonian's first law* (Teghtsoonian, 1971, 1973), the exponent β in the psychophysical power law is inversely proportional to the dynamic range of the continuum (i.e., the log of the physical magnitude ratio of the maximum endurable over the minimum perceptible stimulus). As $\beta = \partial \log \Psi/\partial \log \Phi$, the law implies that the subjective ranges for all continua are equal, which is a partial restatement of *Borg's range model* (Borg, 1962). *Teghtsoonian's second law* (Teghtsoonian, 1971) states (in a simplified formulation, which fits his data slightly better) that the product of β and the Weber fraction $\Delta\Phi/\Phi$ is a constant. This constant is close to 0.03 for a number of continua, which (as $\partial\Psi/\Psi = \beta\,\partial\Phi/\Phi$) led Teghtsoonian (1971) to propose that 0.03 is "the" constant $\Delta\Psi/\Psi$ in Ekman's law.

As $\Delta\Psi$ corresponds to half the interquartile distance in the d distribution, and thus to $0.6745\,\sigma_d$, σ_d/Ψ can be estimated as $0.03/0.6745 = 0.045$. The coefficient of variation in magnitude estimation typically is in the order of 0.20 (Eisler, 1962). Using this figure as an estimate of σ_Ψ/Ψ, we obtain $\sigma_d/\sigma_\Psi = 0.045/0.20 = 0.225$. Computer explorations (setting $\sigma_{\Psi 1} = \sigma_{\Psi 2} = \sigma_\Psi$) reveal optimized values of σ_d/σ_Ψ in this vicinity in, for example, the conditions in Table 1, series 2.

Range and Precision

Teghtsoonian's laws say that the power-law exponent β is determined by the equalization of subjective ranges (law 1) and of subjective variabilities (law 2). The laws may be reconciled as follows. The matching or comparing of two stimuli on different continua but with about equal sensation magnitudes will be most efficient, both for sensitivity to disequilibrium and for freedom from systematic errors caused by the stimulus weighting, if their subjective variabilities are equal. Also, for a match to be possible for every stimulus, all subjective ranges should be about equal. Both these objectives are fulfilled if sensory variability, as indicated by the Weber fraction, is proportional to the dynamic range, which Teghtsoonian's laws jointly predict.

Why is sensory variability so different between continua? One possibility may be the following. Assume that the sensory magnitude Ψ is proportional to the mean impulse frequency during a fixed time interval in J similar, independent channels (Luce & Green, 1972). If J is approximately constant over senses, then a wide dynamic range may need to be partitioned into overlapping subranges, each with its set of specialized channels. As the dynamic range increases, the number of subranges should increase, and their overlap decrease, so that the average number of active channels J goes down, and the variability (the standard error of the mean frequency), which is inversely proportional to $J^{1/2}$, goes up.

Conclusion

I have here collected some observations, conclusions, and suggestions on what happens when we compare two stimuli. Hopefully, the reader is willing to agree that the study of this subject can reveal interesting and important facts about the way we obtain information on our environment. The effects I have discussed are not confined to the laboratory. For example, they may play a great role in everyday comparing of esthetic and social stimuli. The pioneer work in this field remains to be done.

References

Birch, H.G., Belmont, I. & Karp, E. (1965). The prolongation of inhibition in brain-damaged patients. *Cortex, 1,* 397–409.

Borg, G. (1962). Physical performance and perceived exertion. *Studia Psychologica et Paedagogica. Series altera. Investigationes XI.* Lund, Sweden: Gleerup.

Eisler, H. (1962). Empirical test of a model relating magnitude and category scales. *Scandinavian Journal of Psychology, 3,* 88–96.

Eisler, H. (1965). The connection between magnitude and discrimination scales and direct and indirect scaling methods. *Psychometrika, 30,* 271–289.

Ekman, G. (1956). Discriminal sensitivity on the subjective continuum. *Acta Psychologica (Amsterdam), 12,* 233–243.

Ekman, G. (1959). Weber's law and related functions. *Journal of Psychology, 47,* 343–352.

Ekman, G. (1964). Is the power law a special case of Fechner's law? *Perceptual and Motor Skills, 19,* 730.

Fechner, G.T. (1860). *Elemente der Psychophysik [Elements of psychophysics].* Leipzig, Germany: Breitkopf & Härtel.

Gulliksen, H. (1958). Comparatal dispersion, a measure of accuracy of judgment. *Psychometrika, 23,* 137–150.

Hake, H.W., Rodwan, A. & Weintraub D. (1966). Noise reduction in perception. In K.R. Hammond (Ed.), *The psychology of Egon Brunswik* (pp. 277–316). New York: Holt.

Hellström, Å. (1977). Time errors are perceptual. An experimental investigation of duration and a quantitative successive-comparison model. *Psychological Research, 39,* 345–388.

Hellström, Å. (1978). Factors producing and factors not producing time errors: an experiment with loudness comparisons. *Perception & Psychophysics, 23,* 433–444.

Hellström, Å. (1979). Time errors and differential sensation weighting. *Journal of Experimental Psychology: Human Perception and Performance, 5,* 460–477.

Hellström, Å. (1985). The time-order error and its relatives: mirrors of cognitive processes in comparing. *Psychologica Bulletin, 97,* 35–61.
Hellström, Å. (1986a). Sensation weighting in comparing: a tool for optimizing discrimination. In B. Berglund, U. Berglund & R. Teghtsoonian (Eds.), *Fechner day 86* (pp. 89–94). Stockholm: International Society for Psychophysics.
Hellström, Å. (1986b). Sensitivity optimization in psychophysics. In B. Berglund, U. Berglund & R. Teghtsoonian (Eds.), *Fechner day 86* (pp. 95–100). Stockholm: International Society for Psychophysics.
Hellström, Å., Åslund, U. & Almkvist, O. (1985). Comparing computer-presented successive stimuli: indications of cognitive malfunctioning in solvent-exposed workers. In *Neurobehavioral methods in occupational and environmental health. Extended abstracts from the second international symposium, Copenhagen, 6–9 August 1985 (Environmental Health,* Document 3, p. 115–119). Copenhagen: WHO.
Hellström, Å., Forssell, L. G., & Fernaeus, S. E. (1989). Early stages of late onset Alzheimer's disease. V. Psychometric evaluation of perceptual/cognitive processes. *Acta Neurologica Scandinavica, 79,* (Suppl. 21), 87–92.
Hursch, C.J., Hammond, K. & Hursch, J. (1964). Some methodological considerations in multiple-cue probability studies. *Psychological Review, 71,* 42–60.
Köhler, W. (1923). Zur Analyse des Sukzessivvergleichs und der Zeitfehler [On the analysis of successive comparison and time-order errors]. *Psychologische Forschung, 4,* 115–175.
Luce, R.D. & Green, D.M. (1972). A neural timing theory for response times and the psychophysics of intensity. *Psychological Review, 79,* 14–57.
MacKay, D.M. (1963). Psychophysics of perceived intensity: a theoretical basis for Fechner's and Stevens' laws. *Science, 139,* 1213–1216.
Michels, W.C. & Helson, H. (1954). A quantitative theory of time-order effects. *American Journal of Psychology, 67,* 327–334.
Needham, J.G. (1935). The effect of the time-interval upon the time-error at different intensive levels. *Journal of Experimental Psychology, 18,* 530–543.
Needham, E.C. & Black, H.W. (1970). The relative ability of aphasic patients to judge the duration and intensity of pure tones. *Journal of Speech and Hearing Research, 13,* 725–730.
Postman, L. & Tolman, E.C. (1959). Brunswik's probabilistic functionalism. In S. Koch (Ed.), *Psychology: A study of a science* (Vol. 1, pp. 502–564). New York: McGraw-Hill.
Restle, F. (1961). *Psychology of judgment and choice.* New York: Wiley.
Stevens, S.S. (1957). On the psychophysical law. *Psychological Review, 64,* 153–181.
Teghtsoonian, R. (1971). On the exponents in Stevens' law and the constant in Ekman's law. *Psychological Review, 78,* 327–330.
Teghtsoonian, R. (1973). Range effects in psychophysical scaling and a revision of Stevens' law. *American Journal of Psychology, 86,* 3–27.
Van Allen, M.W., Benton, A.L. & Gordon, M.C. (1966). Temporal discrimination in brain-damaged patients. *Neuropsychologia, 4,* 159–167.
Yilmaz, H. (1967). Perceptual invariance and the psychophysical law. *Perception & Psychophysics, 2,* 533–538.

Human Self-Assessment Process Theory: An Eight-Factor Model of Human Performance and Learning; and Everyman's Causation

DARWIN P. HUNT and MICHELLE R. SAMS

Abstract

This chapter focuses and elaborates on an eight-factor model of the processes by which people make decisions and select responses. The model is part of a larger conceptual framework (called human self-assessment process theory) which describes how people might make self-assessments of the state of their own subsystems and how these self-assessments affect their decisions, and the selection and execution of responses. We think the topic, because of its potential use for making predictions and in understanding how and why people behave as they do, is deserving of more systematic consideration than it has received.

An intervening variable approach is employed to explore the development and the effects of associations:

1. Between responses and particular consequences (later events) – called response-consequence associations
2. Between perceived situations and later events – called situation-consequence associations
3. Between perceived situations and responses – called situation-response associations

These associations involve what may be called cause and effect relationships. The condition(s) which a person identifies as the cause(s) is affected by many factors such as (a) the person's educational and experiential background, (b) the gains and costs related to errors of causal inference, (c) the conditions over which the person has control – called the principle of relativity of causes, (d) human biases exhibited when diagnostic conclusions based on past hypotheses-symptom data are more sure than is warranted, and (e) the necessity that an informational relation exist between two events before a person will form a cause-effect association between them; repeated observation of regularity between events is neither sufficient nor necessary for a person to infer a cause–effect relation.

Introduction

The assessments which a person makes of *how sure* he or she is that they can answer some question correctly, perform some action properly, attain some goal acceptably, etc., and the effects of these self-assessments on the decisions and actions of that person is the topic of central interest in this discussion. We propose that the quality and quantity of the performance and learning of a person in everyday life in work and play is importantly affected by these self-assessments (Hunt, 1986).

Our comments also apply to self-assessments of an answer or a response which has already been executed by a person, but for which no knowledge of results or feedback about its correctness has yet been received. Of course, the execution of the response or action can be expected to influence reactively the self-assessments, e.g., additional stimuli are produced by executing the response which may result in a more accurate self-assessment. For example, one may feel unsure of the combination of a lock until one physically spins the combination knob.

To perform effectively and efficiently, it is not sufficient for a person to possess the knowledge and skills to perform some task. It is also necessary that the level of sureness concerning the possession of the knowledge and skills (and perhaps the ability to retrieve it from memory and to execute the essential responses) be high enough that, when the person is confronted with a choice of whether to engage in the task, he or she will decide to do so.

In addition, the latency, vigor, and smoothness with which the responses are executed are related to the accuracy of a person's self-assessments. For example, when people were learning the names of hand tools it was found that answers about which the learner was *sure* and which in fact were correct were executed approximately 1.5 s faster than were *sure*-but-wrong answers (Hunt, 1984).

Experimental findings from our own laboratory which are relevant to, i.e., support or do not support, the above conclusions and which provide additional discussions of self-assessment responding are reported elsewhere (Crouse, 1980; Fernandez, 1982; Hughes, 1975; Hunt, 1976, 1977, 1978, 1981, 1982a, 1982b, 1984, 1986; Sams, 1986; Shurtleff, 1987).

A conceptual framework to interpret the findings regarding human self-assessment responding is presented in Fig. 1 and has been described elsewhere (Hunt, 1984, 1986). For those who are congenial to following arrows from one block to another, we hope that they will do so in Fig. 1 and will be enlightened by the exercise. For those who are antagonistic to block-arrow diagrams, simply notice that in the upper left part of Fig. 1 there is outlined by a fuzzy line a portion labeled "Eight-factor model", which will be the focus of the remainder of this chapter.

The eight-factor model, shown in Fig. 2, and the general framework are both considered to represent a moment in time; they move through successive moments in time. As a result of interactions with events and objects in the

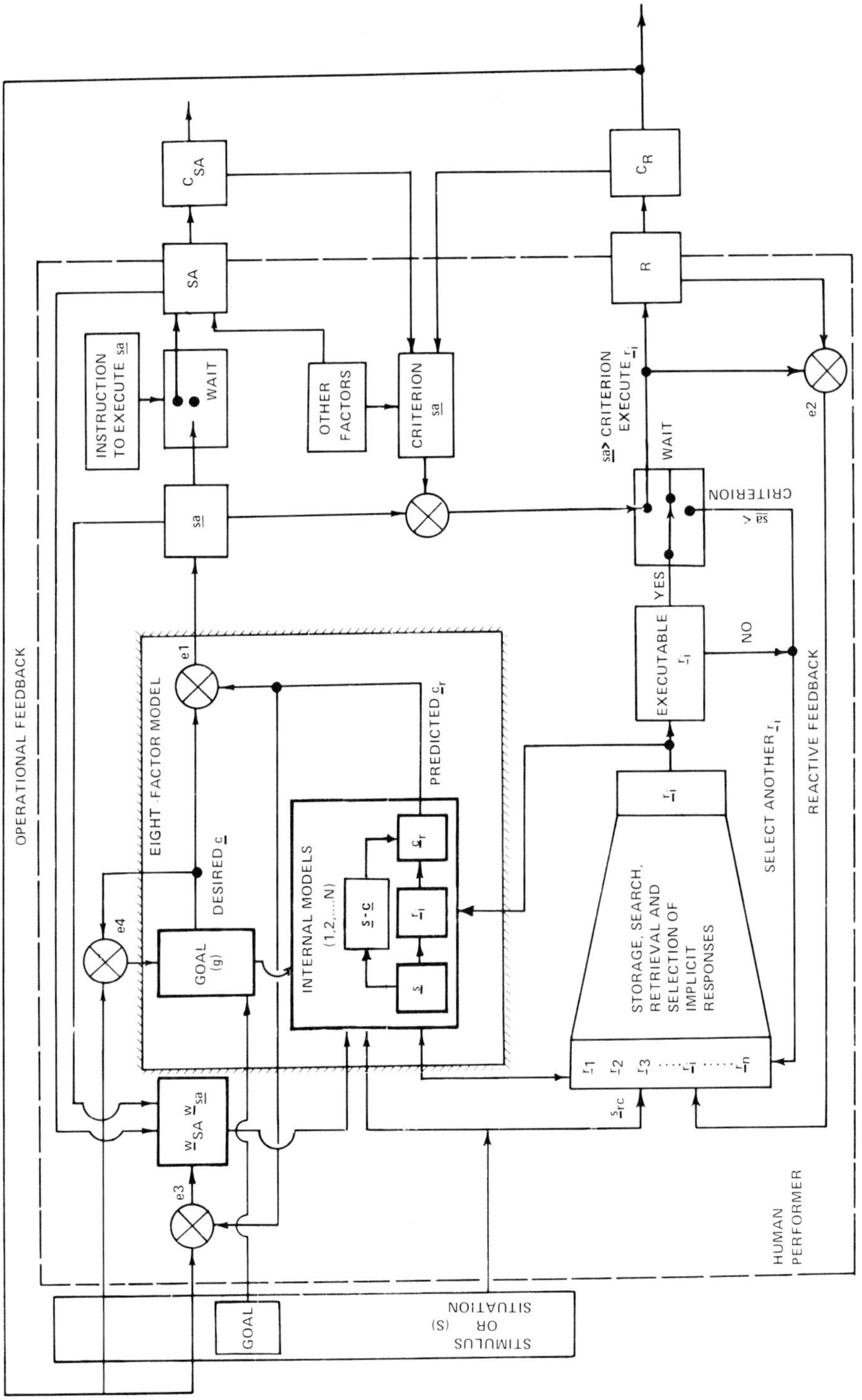

Fig. 1. A framework to conceptualize human performance and learning and the eight-factor model

real-world environment, or more precisely the person's perception of the events and objects, the various factors are modified. This could entail learning to perceive the situation more accurately and/or learning an $r - c_r$ association, i.e., a relationship between a response and the consequences expected to be produced by it, etc.

The self-assessments of a person are assumed to affect the locus and the amount of learning that takes place. After a response is executed and the feedback consequences are received by the person (e.g., "Your answer is correct"), the prior self-assessment influences the extent to which the factors in the model are modified. The influence of self-assessments on learning, i.e., on modifications of subsequent behavior, has been experimentally demonstrated by findings that learning was faster and greater for learners who made self-assessment responses than for those learners who did not (Hunt, 1978, 1982b; Sams, 1986).

Eight-Factor Model

The reluctance to construct this eight-factor model (because of our limited knowledge of how the human processes operate and our uncertainty as to how to incorporate what we do know) was overcome by the need for an integrated way of thinking about human self-assessment, performance and learning in the everyday life of people. The aim of the model seems consistent with Skinner's (1987) comment that we need to put ". . . Humpty Dumpty together again by studying relatively complete episodes, each with a history of reinforcement, a current setting, a response, and a reinforcing consequence" (p. 782). We will elaborate on these and other components of episodes as viewed within the eight-factor model.

The model was designed to be consistent with our knowledge about the various psychological processes of the human subsystems (perception, learning, decision-making, and motor/muscular movement). The eight factors are considered to be intervening variables or theoretical concepts which have been temporarily adopted to help understand, summarize, synthesize, interpret, and predict (human) behavior and performance. Of course, the scientific task remains for us to elaborate the details of how independent variables are related to the various factors and how the various factors combine to predict behavior.

The factors which we have chosen to include in the model are intended to:

1. Imply that each factor represents a class of potential human performance competencies or deficiencies, which may be measured and identified, and which can be modified by various means such as training and redesign of the interfaces between the person and machines/environment
2. Provide a set of factors to be employed to describe human tasks in a form which will be helpful in the selection of those media and methods of instruc-

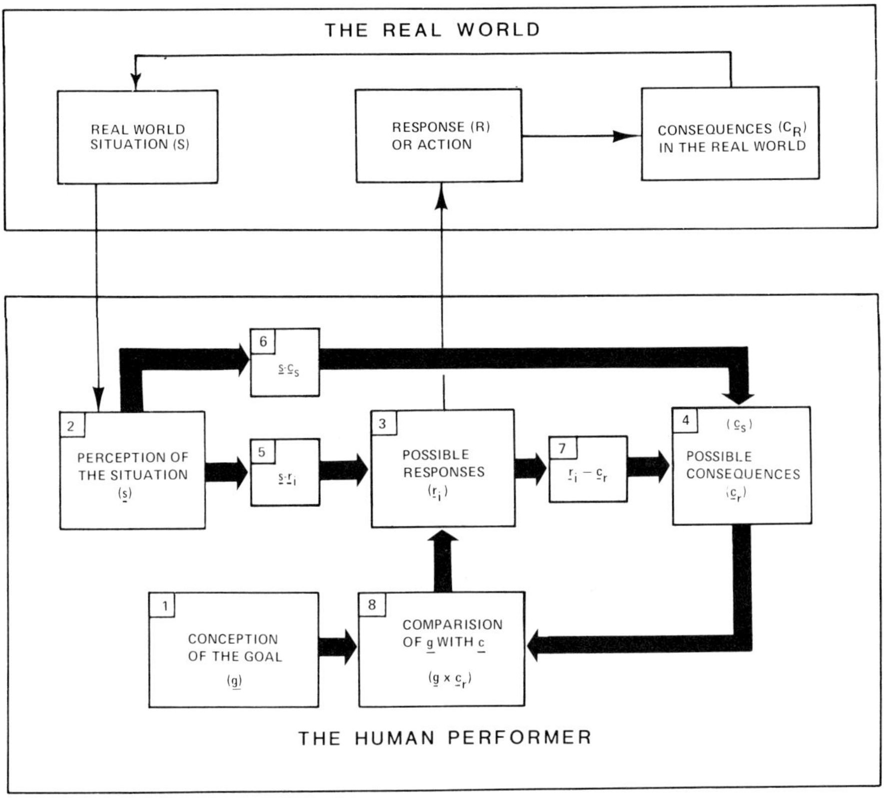

Fig. 2. An eight-factor model of human performance and learning. (Based on Bobrow, 1975; Hunt, 1984; and many others)

tion, learning, or training which will be most effective, economical, and efficient
3. Provide a conceptual framework and model of human processes which are amenable to mathematical and computer simulation and manipulation, and
4. Provide a model which will encourage the development of taxonomies of task components (situations, goals, responses, etc.) and tasks

The eight factors, with brief comments about each, are:

1. Goal (g). It is assumed that a person behaves in accordance with purposes and goals. Here we will not distinguish between purposes and goals – they are used interchangeably. The way in which the person interprets, defines, and conceptualizes various characteristics of the goal (such as the concept of correct performance, error tolerances, allowable time, etc.) provide internal standards against which to assess, select and execute actions.

2. Perception of the situation (s). An existing situation (S) is perceived (*s*). Effective performance requires that the situation be detected and identified with a speed and accuracy commensurate with the performance required by the goal.

A real-life situation is considered to consist of a set of conditions which combined produce, at a later moment in time, another situation. A condition may be considered to be a state of some environmental dimension, such as the temperature of some object. The perception (*s*) of a situation may be defined as the extraction of information from the set of conditions by the person.

We consider that a person has available cues, either directly through observation of the situation or indirectly by means of informational displays and controller devices of machines. These cues provide data to the person about the state of selected environmental dimensions – which, in turn, can allow the person to draw inferences about, i.e., perceive, the situation. The environment may contain human as well as nonhuman objects, events, and relationships.

3. Response repertoire (r). A repertoire of responses and actions (a sequence of responses) which the person is capable of executing must be sufficiently well learned and practiced so that they can be selected, covertly assessed, and executed when required. These responses may produce or prevent one or more conditions in a later situation.

The complexity of the response will be partly determined by the kind of goal which is involved in the task being performed by the person. For example, the goal of building a house requires a more complex response sequence than does buying a doorknob. Some very complex response sequences which are well practiced and well learned, such as an adult tying a shoe string or various movements of performers on musical instruments, sports, dance, etc., may be accomplished with relatively little attention and effort. Such complex response sequences become in the language of Kelley (1968), inner loop responses which are more-or-less automatically executed when the outer loop selection of an action is made.

4. Consequences (c). This factor is concerned with the conceptualization and enumeration of possible outcomes or consequences of responses. Because a person must compare these anticipated consequences with the goal, it seems reasonable that whatever framework is used by the person to conceptualize or construe the goal is also the same framework employed to conceptualize the consequences. Furthermore, it seems parsimonious to propose that a person conceptualizes possible outcomes in terms of various conditions, i.e., states of a set of dimensions, much like the manner proposed for the perception of a situation.

5. Situation–response (s–r) associations. The person possesses associations, i.e., connections, understandable relations, etc., between perceived situations and responses which affect the retrieval of responses from memory. It is a long-held view in psychology that the occurrence of a particular stimulus tends to evoke particular responses ($r_1, r_2, \ldots, r_n$) with a relative frequency corresponding to the strength of the $s-r$ associative bond. Part of the training process involves the modification of these $s-r$ associations.

6. Situation–consequence ($s-c_s$) associations. By observing the regularities in his or her environment, a person acquires associations between situational (other people and natural) events which occur and the later occurrence of other events (consequences). The linking together of such observed events allows the person to behave as if there exists a causal sequence. These acquired associations among events provides a means for a person to control indirectly events far along the chain.

It is assumed that people infer such causal relationships and make decisions and execute actions in accordance with them. Bolles (1972) has developed in great detail the notion that learning consists of the acquisition of expectancies, including $s-c_s$ expectancies, i.e., that particular environmental events (stimuli, conditions, or situations) predict particular other events.

The methods employed by people to make inferences to identify the cause-effect conditions and relations depend upon their educational and experiential background, the gains and losses which may be related to errors of causal inferences, the person's values, etc. Scientists may employ a formal step-by-step approach (called the scientific method) to identify causes; but others may have developed their own private, less rigorous methods for observing regularities in their own daily lives.

7. Response–consequence ($r-c_r$) associations. By learning and experience, particular consequences are associated with the responses which the person believes to have produced or prevented them either directly or indirectly. After a response is executed, the feedback concerning the real-life consequences which occur plays a vital role in modifying these $r-c_r$ associations.

The direct consequences of a person's responses may be called first-level outcomes (Hellriegel, Slocum, & Woodman, 1983), which correspond to the $r-c_r$ associations. The $s-c_s$ associations are related to second-level outcomes which a person expects to occur as a result of the occurrence of first-level outcomes. For example, an employee may begin to play golf with the boss observing that others who do so are promoted quickly.

Rescorla (1987) recently analyzed goal-directed behavior with regard to the role of the response, the contingent reinforcer, and the stimulus in the presence of which that contingency is arranged. He concludes that an association between a response and its reinforcer is of basic importance in the acquisition and maintenance of instrumental behavior.

His discussion suggests that the perception of a particular situation may set the occasion for the development of an $r-c_r$ association between the response and particular consequences. Extending this notion, we suggest that the $r-c_r$

associations may be more in the form of $(r/s) - c_r$ associations. Thus, *s*, the perception of situation, may participate in the selection of the response in this higher order fashion.

8. Goal-by-consequence comparison ($g \times c$). The anticipated consequences (c) are covertly compared with the goal (g) to determine whether the response being tested should be executed or rejected. If there is a sufficiently close match, then the response is executed; otherwise the response is rejected and another response is retrieved from memory and covertly tested. It is this $g \times c$ comparison which is assumed to produce a level of sureness, or self-assessment (*sa* in Fig. 1) of a person. This *sa*, when compared to a *criterion sa,* partly determines whether a particular response will be executed and also affects the latency, smoothness, etc., with which it is executed.

An Intervening Variable Approach and Mental Models

You may observe that the factors in the model are sometimes discussed as if they were components in a mental model which a person constructs cognitively (in the mind) and uses to organize the world around them and to execute actions (Jagacinski & Miller, 1978), rather than as scientific concepts in a human peformance theory. We find this apparent inconsistency unsatisfying. However, we were unable to force our terminology into only the scientific concept mold and decided to simply accept, for the purposes of this paper, the inconsistent usage.

There is a difference between the mental model approach and our approach, which may be called an intervening variable approach. Gentner and Stevens (1983) indicate that mental model research is an effort to capture naturalistic human knowledge and that their representations of knowledge are based on formalisms arising from artificial intelligence research, e.g., production rules, constraint networks, etc.

In contrast, we see ourselves as attempting to describe the human psychological processes in terms of intervening variables (Chaplin & Krawiec, 1968). To us this means describing the functional characteristics of the components, and the channels and operations by which the components interact. The adequacy of the conceptual (and eventually mathematical) description may then be evaluated in terms of the extent to which deductions (outputs) from the model of the human processes agree with experimental data. When disagreement is found, a component (factor or relationship) may be revised or a new component may be added.

It is hoped that this approach will later produce formal rules or postulates relating the fundamental intervening variables to each other and to their anchors in environmental events. For example, a postulate might be:

Postulate 8. Selection of a covertly considered response (r_i). For a covertly considered response (r_i) to be selected, the level of the *sa* must be greater than or equal to the *criterion sa* (C_{sa}). That is,

$r_i = sa - C_{sa}$, where if $r_i \geq 0$, then execute (R) the response r_i;
if $r_i < 0$, then consider alternative responses (r_i) or modify the conception of the goal (g).

A proper set of postulates, together with adequately defined terms, should generate testable deductions. In summary, our model is a set of anchored factors beginning with the antecedent conditions (input), moving through the intervening variables to the response (output).

The focus is on the processes that produce the output, and not on mimicking the output. Among ourselves, we call this a "Keplerian" approach to modeling and computer simulation. An alternative might be called an "epicycle" approach – suggesting a pre-Kepler theory of planetary motion which allowed the location and movements of planets to be predicted quite well. The distinction is that, even though predictions about the output might be made as accurately by an epicycle model as by a Keplerian model, the processes by which the Keplerian model generates the predictions is more in keeping with the fundamental relations and operations of objects and events in real life.

Everyman's Causation

As may be apparent, the processes by which a person detects, identifies, stores in memory for future use, retrieves, and utilizes the perceived relations among situations, responses, consequences, and goals involve what may be called cause and effect relationships. Our main interest, as applied experimental and ergonomical psychologists, in the concept of causation is in trying to describe the psychological processes by which a person (a) inferentially identifies relationships among situations, responses, and consequences, (b) learns the relationships as $s-r$, $s-c_s$, and $r-c_r$ associations, and (c) uses these remembered associations in everyday life.

Here we do not address how people ought to judge causes and effects. We take the position that most people behave *as if* there are relatively simple causal events which produce later effect events. We will focus on the fact that people perceive and draw inferences about what antecedent events or conditions will cause or have caused what later events, and on how their decisions and actions are affected by these perceptions.

There are many definitions of the term *cause*. Two (of three) provided by Collingwood (1976) which relate to the eight-factor model will be emphasized. We will refer to them as the "practical" sense and the "theoretical" sense of cause.

The first definition relates $r-c_r$ associations to the practical sense of cause as "... an event or state of things which it is in our power to produce or prevent, and by producing or preventing which, we can produce or prevent that whose cause it is said to be" (Collingwood, 1976, p. 180). It is important to note that, in the practical sense, the cause is that which is immediately under human control, i.e., which can be produced or prevented by a person, and the effect either (a) can be produced directly (first-level outcome) or (b) can be produced indirectly (second-level outcome) by the person because of the relation in which it (the second-level outcome) stands to an intermediate causal condition. For example, a person can directly produce a state of things for an electric light switch called, "The switch is at the ON position". Turning the switch, under certain conditions, is a person's means of producing a further state of things called, "The light is lit".

The second definition of cause relates $s-c_s$ associations to the theoretical notion that natural events may "... happen independently of human will but not independently of each other: causation being the name by which this dependence is designated" (Collingwood, 1976, p. 171). This theoretical notion of cause refers to a set of conditions that bears a relation such that (a) if the cause happens or exists under certain conditions the effect also must happen, (b) the effect cannot happen unless the cause happens, and (c) the cause is prior to the effect.

Having available particular $s-c_s$ associations is necessary if a person is to be able to select actions in accordance with the goals which he or she is seeking. The notion of an $s-c_s$ relationship is that particular s events in the person's environment, which a person may not have produced, are followed more-or-less regularly by particular c events. A person may observe these regularities, infer a cause–effect relation, and learn $s-c_s$ associations. (Such uncontrolled environmental conditions may also serve to constrain the kinds of responses available to a person.)

Much of our formal education and learning in the sciences, including the behavioral and social sciences, is of this kind of value, i.e., it provides means for an individual to control directly and indirectly objects, events, and relationships. Education provides us with knowledge of $s-c_s$ causal relationships which, if we can identify a causal factor which we can produce or prevent, then we can select a response to bring about an effect (consequence) which matches our desired goals. The $s-c_s$ associations which we have stored in our memory for use when needed are acquired through our individual experiences, personal observations of regularities and learning, as well as through formal education.

Usually, the associations are not formed between two events simply because they are contiguous in time, but there must exist some informational relation between them as well (Rescorla, 1987). A causal relation might also be perceived without repeated observations of regularity between two events. For example, an event may be identified as a cause if it is the *only* change (believed to have been) introduced into the situation immediately before some effect event occurs (Ducasse, 1976).

Indeed, it may be more reasonable to view the regular succession of two events as raising the question, rather than providing the answer, as to whether one causes the other, or whether some antecedent third factor caused both, or whether the conjunction of the two is simply accidental (Ducasse, 1976). Two clocks striking the hours, one of which always strikes immediately before the other is likely due to a third factor – the particular time of day and clock characteristics – rather than the first clock causing the second to strike a specific number of times.

Many times there are biases and errors made when inferring causal relationships. These inferences include prediction (proceeding from cause to effect) and diagnosis (proceeding from effect to cause). From the person's perspective, there is a set of complex antecedent conditions, one (or a small number) of which is identified as the *cause* and some of which may simply be called the *conditions* (some called predisposing conditions and others nonrelevant). At later moments in time there are successive sets of complex conditions which occur; one condition (or more) in each set may be identified as the effect. From a practical point of view, a person is usually interested in the conditions which he or she is able to produce or prevent when desired, e.g., to attain particular goals.

Eddy (1982) has identified cause–effect biases in medical diagnoses. There may exist a high probability that an individual who has a particular disease will show a certain symptom, yet there may be a low probability that one who shows the symptom will have the disease. Eddy points out that, in such situations, physicians tend to be overconfident in diagnosing the disease in a patient who shows the symptom.

Indeed, it has been proposed (Collingwood, 1976) that the particular condition which one labels the cause is that condition which one is able to produce or prevent. This principle of "relativity of causes" suggests that the condition which is identified as the cause by one person may be different for another person depending upon the conditions in the environment over which each person has control, i.e., the conditions which the person can produce or prevent. Thus, as in our medical example, the same ill patient could be diagnosed as needing surgery by a surgeon or needing a change in diet by a nutritionist.

Of course, it is a difficult task for the person to distinguish between those later conditions which were produced or prevented by one's own response, directly or indirectly, and those later conditions which were not, i.e., conditions which were produced by natural events or by other people. We view this within the eight-factor model as the distinction between $r-c_r$ associations and $s-c_s$ associations.

The distinction is important because the ability to produce or prevent particular conditions provides the means of control for a person. Indeed, the extent of control one has might be viewed as the likelihood that particular events will occur (a) if the person makes a particular response compared to (b) if the person does not make that particular response.

However, it should be noted that there is lacking in everyday life a way for a person to compare what would have happened if the response had not been executed or if a different response had been executed, i.e., life has no control group. Thus, when two or more conditions occur together a person may infer that the effect or outcome was due to one particular condition, when it was actually produced or prevented by one or more other conditions.

In the end, the answer to the question of whether it is worthwhile to relate the concepts of associations, causation, the eight-factor model, and human self-assessment process theory will be "... in terms of the range of phenomena to which [they] can be profitably applied" (Klein, 1976, p. 171).

Acknowledgments. We are pleased to help celebrate – by providing the present chapter – the numerous, multidisciplinary, continual, and high-quality contributions of Gunnar Borg (e.g., Borg & Ottoson, 1986). Almost all research scientists are able to formulate questions or hypotheses which are answerable. However, the research task which is most demanding of creative thought is the formulation of research questions which are important as well as answerable. The work of Professor Borg has exhibited such creativity and importance by contributing both to the theoretical/conceptual thinking in the domain of physical exertion and psychophysical scaling and to the practical application by the development of his ratings of perceived exertion (RPE) scales. It is also a timely invitation because it provides an opportunity to write about a topic of great current interest to us.

References

Bobrow, D. G. (1975). Dimensions of representation. In D. G. Bobrow & A. Collins (Eds.), *Representation and understanding: Studies in cognitive science*. New York: Academic.

Bolles, R. C. (1972). Reinforcement, expectancy and learning. *Psychological Review, 79* (5), 394–409.

Borg, G. & Ottoson, D. (1986). *The perception of exertion in physical work*. London: Macmillan.

Chaplin, J. P. & Krawiec, T. S. (1968). *Systems and theories of psychology*. New York: Holt, Rinehart and Winston.

Collingwood, R. G. (1976). Causation. In M. Brand (Ed.), *The nature of causation*. Urbana, IL: University of Illinois Press.

Crouse, R. W. (1980). *The novelty of human self-assessment: Implications for learning and training*. Unpublished master's thesis, New Mexico State University, Las Cruces.

Ducasse, C. J. (1976). Causality: Critique of Hume's analysis. In M. Brand (Ed.), *The nature of causation*. Urbana, IL: University of Illinois Press.

Eddy, D. M. (1982). Probabilistic reasoning in clinical medicine: Problems and opportunities. In D. Kahneman, P. Slovic & A. Tversky (Eds.), *Judgment under uncertainty: Heuristics and biases*. New York: Cambridge University Press.

Fernandez, J. (1982). *Self-assessment and decreasing stimulus duration*. Unpublished master's thesis, New Mexico State University, Las Cruces.

Gentner D. & Stevens A. L. (1983). *Mental models*. Hillsdale, NJ: Erlbaum.

Hellriegel, D., Slocum, J. W. Jr. & Woodman, R. W. (1983). *Organizational behavior* (3rd ed.). New York: West Publishing.

Hughes, C. R. (1975). *The relationship between self-assessment and response choice during learning*. Unpublished master's thesis, New Mexico State University, Las Cruces.

Hunt, D. P. (1976). *The human self-assessment process: Study I. Its accuracy, effects on acquisition and a tentative model* (An Interim Report). Alexandria, VA: U.S. Army Research Institute for the Behavioral and Social Sciences.

Hunt, D. P. (1977). *The human self-assessment process: Study II. The effects of the number of self-assessment categories on acquisition* (An Interim Report). Alexandria, VA: U.S. Army Research Institute for the Behavioral and Social Sciences.

Hunt, D. P. (1978). *A human self-assessment process* (A final report). Alexandria, VA: U.S. Army Research Institute for the Behavioral and Social Sciences.

Hunt, D.P. (1981). *A human self-assessment process* (Tech. Rep. No. 466). Alexandria, VA: U.S. Army Research Institute for the Behavioral and Social Sciences.

Hunt, D.P. (1982a). *Human self-assessment responding: A consideration of the motor component and of the speed and accuracy of self-assessment responding* (An Interim Report). Alexandria, VA: U.S. Army Research Institute for the Behavioral and Social Sciences.

Hunt, D.P. (1982b). Effects of human self-assessment responding on learning. *Journal of Applied Psychology, 67,* 75–82.

Hunt, D.P. (1984). *Human self-assessment and its implications for human training and performance.* Alexandria, VA: U.S. Army Research Institute for the Behavioral and Social Sciences.

Hunt, D.P. (1986). The ability of people to make self-assessments of the state of their own subsystems: Implications and the process. In G. Borg & D. Ottoson (Eds.), *The perception of exertion in physical work.* London: Macmillan.

Jagacinski, R.J. & Miller, R.A. (1978). Describing the human operator's internal model of a dynamic system. *Human Factors, 20,* 425–433.

Kelley, C.R. (1968). *Manual and automatic control: A theory of manual control and its application to manual and to automatic systems.* New York: Wiley.

Klein, R.M. (1976). Attention and movement. In G.E. Stelmach (Ed.), *Motor control.* New York: Academic.

Rescorla, R.A. (1987). A Pavlovian analysis of goal-directed behavior. *American Psychologist, 42,* 119–129.

Sams, M. (1986). *Effects of testing methods and self-assessment responding on observers and performers.* Unpublished master's thesis, New Mexico State University, Las Cruces.

Shurtleff, M. (1987). *Training novice troubleshooters to use the half-split rule.* Unpublished master's thesis, New Mexico State University, Las Cruces.

Skinner, B.F. (1987). Whatever happened to psychology as the science of behavior. *American Psychologist, 42,* 780–786.

For Hedgehogs and Foxes: Individual Differences in the Perception of Cross-Modal Similarity *

LAWRENCE E. MARKS

Abstract

The psychophysical study of human perception can satisfy both Platonistic hedgehogs and Aristotelian foxes – for psychophysics blends general, even universal scientific principles with individual variation and particularity (for example, Weber's law of intensity discrimination, Stevens' law of intensity scaling, and Borg's principle of maximal intensity perception). Notable in this regard is the realm of perceptual similarity: studies of cross-modal similarity in general, and studies of synesthesia in particular, reveal unifying principles in combination with salient individuality. People appreciate systematic equivalences between perceptual qualities of experience from different modalities – equivalences, say, between bright colors and loud sounds or between bright colors and high-pitched sounds; yet at the same time people establish unique and sometimes idiosyncratic correspondences between perceptual qualities of the different sensory modalities.

To those of us who contemplate the rich complexities of psychology, the discipline of psychophysics exerts a great attraction, for psychophysics promises to create order out of a blooming, buzzing confusion by uncovering general, even universal, rules of mind and behavior. Not many branches of psychology have succeeded, over the course of the century or so since it established itself as a science, in harvesting general "laws." Indeed, when I try to conjure up the names of laws of psychological science, exempting those that lie within psychophysics, I tend to find a meager handful of items – some, like the Yerkes-Dodson law, have aged poorly, fading over the years in a cloud of increasing obscurity; others, like the law of disuse or the law of effect, comprise either statements of limited application or validity or statements of definition.

Practitioners of psychophysics, by way of contrast, over the years have reaped a cornucopia of principles of sensory and perceptual function, establishing useful quantitative rules under such eponymous titles as Weber's law, Stevens' law, Emmert's law, the Ferry-Porter law, Bloch's law, and Ricco's law. Not every one of these laws is universal, to be sure; but certainly they

---

* Preparation of this chapter was supported by U.S. National Institutes of Health grants NS-16993 and NS-21326.

are wide-ranging in scope and nutritious to the scientific soul that feeds on nomothetic principles.

Much of this attraction that universals generate, much of their appeal, stems from the power, the beauty, and, ultimately, the simplicity that they engender. For a single general law or rule can govern dozens or even millions of specific cases, providing a looking glass through which scientists can observe and interpret a Blakean world in a grain of sand. Yet at the same time even the most ardent of epistemological monists must confess, and perhaps puzzle over, the fact that every human being is unique – and indeed can relish the fact that individual minds, behaviors, and experiences, even when focused through the restricted lens of a psychophysical microscope, are diverse and sometimes idiosyncratic.

The seeming contrast between the essences of universals and the diversity of particulars has provided a focal point of controversy in philosophy since the times of Plato and Aristotle. It is epitomized in the conflict, popularized by Sir Isaiah Berlin (1953), between "the hedgehog [who] knows one great thing" and "the fox [who] knows many things." The dichotomy between hedgehogs and foxes represents a dispute between those who believe in unity and those who believe in diversity; it is a dispute that continues in philosophy, in social and political thought, and in psychology. What I wish to point out here are a few instances where monism and pluralism meet, thereby providing psychologists with a chance to have their whole cake and eat all of its pieces too.

A few psychophysical laws, at least, accommodate both the universal and the individual: Like certain laws of physics (the universal gas law provides a good example), psychophysical laws can provide overarching principles that are common to all the elements in their domains, yet that leave those elements free to vary in their characteristic fashion.

Let me give two examples: Weber's law of intensity discrimination and Stevens' law of intensity scaling. Weber's law states that just-detectable increments in stimulation are constant in percentage terms. This is the general rule. It says that ΔS, the change in stimulus magnitude that can just be detected, is a constant percentage or proportion k of the magnitude S from which the change is made. Mathematically, $\Delta S = k\,S$. But the actual percentage itself can shift from sense modality to sense modality or even across conditions within a given modality. These are the individual cases. For example, k equals about 3% for detecting changes in line length, but 8% for detecting changes in sound intensity; and k equals about 10% for detecting changes in light intensity when the visual fields being compared are spatially separated, but only about 1% when the visual fields have contiguous borders, thereby invoking processes of spatial contrast.

Stevens' power law states that equal ratios of stimulus intensity produce equal ratios of sensation magnitude – again, this is the general rule. In mathematical form, it says that sensation magnitude M rises as a power function of stimulus intensity S: $M = c\,S^b$. Stevens' law applies to all or virtually all sensory modalities (to vision, hearing, taste, smell, and touch), and often to different dimensions or attributes in a given modality (in vision, for instance, it characterizes the behavior of brightness, saturation, length, and area). But the actual ratio relations, as reflected in the size of the exponent b, depend on the sense modality or perceptual quality in question – again, these are the individual cases. For example, augmenting the physical length of a line tenfold causes a comparable tenfold gain in its perceived length ($b=1$), but similarly augmenting sound intensity tenfold causes only a doubling in the loudness of medium- and high-frequency sounds ($b=0.3$) and a somewhat greater than doubling in the loudness of low-frequency sounds ($b=0.3-0.6$).

Nearly three decades ago, Gunnar Borg (1961, 1962) proposed a rule that has just this kind of power, the power of capturing a universal essence, yet accommodating individual variation. Borg argued that, at least to a first approximation, the perceptual experience of exertion during dynamic exercise (say, bicycling) is the same in all people when they work at their respective maxima. This simple principle serves both to identify what is common to the normal perception of exertion and at the same time to establish a framework for evaluating what is uncommon or unique. So for example when both you and I pedal a bicycle at our respective maxima, your experience in some sense "equals" my experience; in this respect, Borg's principle expresses a universal. Yet at the same time you may be considerably more proficient at cycling or better trained than I, and thus at maximum you may be performing quite a different level of physical work.

If this principle is correct, then, Borg inferred, it is also possible that when people use a given verbal label or expression, such as "weak" or "very intense," to describe their level of exertion, then the corresponding perceptual experiences also will be roughly commensurate. Indeed, experiments using heart rate as an independent measure of physiological strain – a measure that presumably parallels perceived exertion – provided evidence that this inference is correct (Borg, 1972, Marks, Borg, & Ljunggren, 1982).

What are the universals and what are the particulars of sensory and perceptual experience? For a goodly number of years now, I have been exploring an issue that directly addresses this question. The issue is similarity, and in particular, similarity across the senses. While most of us recognize the qualities of sight, hearing, touch, taste, and smell as being distinct and differentiated – hence a commonsense basis for characterizing different sense modalities – there is a small fraction of people who experience their universe synesthetically (Marks, 1975, 1978), which is to say that they find the perceptual qualities of different sense modalities to be intimately connected. In synesthetic perception, music and voices contain not only auditory qualities of timbre, loudness, and pitch but may also contain visual qualities of color, brightness, and shape. To synesthetic individuals, there is a fundamental interconnectedness, a fundamental unity, across the senses.

At a first glance, synesthesia may seem to epitomize the individual and the idiosyncratic, transporting uniqueness of phenomenal experience virtually to solipsism. Every synesthetic person is different; every one is able to describe his or her set of intersensory associations in personal terms, recounting the particular color aroused by each note of the musical scale, by each musical instrument, or by each quality of taste. C-sharp may be green to Ingrid, but violet to Karl; to a composer, the deep voice of a trumpet may be red or blue or yellow; to a swimmer, the taste of brine may be pink or white or brown – in these associations we find a world of intersensory correspondences, many of them explored by symbolist poets in the last century and by futurists in our own.

Yet as I indicated, as individual and idiosyncratic as synesthetic perceptions may be, they are not *merely* individual, they are not *merely* idiosyncratic. Synesthetic perceivers may, and usually do, disagree as to the particular colors evinced by particular musical notes, but nevertheless they may agree – and, again, they usually do – that whatever their hues the tints are darker or brighter as the octave or the musical register is lower or higher. In low octaves a note of A is deep, somber, and dark, whether it appears cloaked in scarlet or navy or brown, while in higher octaves A is clear, light, and bright, whether bright red or bright blue or bright yellow. That is, hue may be personally related to musical pitch, but brightness is universally related to pitch height (see Marks, 1975).

Even those of us who do not experience the world synesthetically can appreciate resemblances between select aspects of auditory experience and select aspects of visual experience. Much as synesthetic perceivers report that low-pitched notes surround themselves in vast, dark colors, and high-pitched notes concentrate themselves in smaller, bright colors, so do nonsynesthetic perceivers – who constitute the vast majority of us – assess in just the same fashion the relation between tones and colors (a relation that may seem to the nonsynesthetic more conceptual than perceptual). In a cross-modality matching task, for example, virtually all subjects will set higher sound frequencies to match greater visual intensities (Marks, 1974, 1978), thereby revealing a universal appreciation of similarity between the dimension of pitch on the one hand and that of brightness on the other. Incidentally, this particular intersensory relation, at least as it applies to synesthesia, was lifted to the status of a "law" back in the last century, both by Bleuler and Lehmann (1881) in their *Helligkeitsgesetz,* and by Flournoy (1893) in his *loi de clarté.*

Not surprisingly, synesthesia and cross-modal perception reveal several other universal principles of intersensory similarity besides pitch-brightness. A second correspondence between hearing and vision involves intensity: loud sounds tend to be perceived as visually bright by synesthetics, and are matched to bright lights by nonsynesthetics, while soft sounds tend to be perceived as dim by synesthetics and to be matched to dim lights by nonsynesthetics (Marks, 1978). So the pairing of loudness and brightness joins that of pitch and brightness as a universal of cross-modal experience. The universality of pitch-brightness similarity and of loudness-brightness similarity, by the way, extends down into childhood – at least to age 4 years, according to results that Marks, Hammeal, and Bornstein (1987) obtained by a cross-modal matching procedure, and maybe even to the first month of life, according to results that Lewkowicz and Turkewitz (1980) obtained by a procedure that measured cross-modal generalization of heart rate.

A rule that says "Brighter lights match louder sounds" can serve as a very nice general principle, but it leaves open the question of individual variation within the structure imposed by the cross-modal correspondence. For now, let us restrict ourselves to variation within the domain of intensity itself, in order to ask, How well do people agree on the cross-modal correspondences of perceived intensity? You and I may agree on *relationships*, that is, we may agree that the brighter of two lights corresponds to the louder of two sounds, more or less independently of their absolute intensity levels, but we may disagree on the absolutes, arguing whether the brightness of the moon viewed on a clear night matches the loudness of ordinary conversation or the loudness of a shouting match. Indeed, one of the first questions we may ask is whether individuals are at all reliable with respect to the absolute levels of their cross-modality judgments.

Individuals surely do differ in the absolute levels of stimuli they judge to match perceptually. Stevens and Marks (1980) used a new method they developed, called magnitude matching, to assess the similarity between the brightness of white light and the loudness of 1000-Hz tone. In magnitude matching, stimuli that differ in physical intensity and that belong to different modalities – in this case, lights varying in luminance and sounds varying in sound pressure level (SPL) – are presented on different trials, while the subjects try to judge all of the stimuli on a common scale. Stevens and Marks used an open magnitude-estimation scale, but judgments may also be made on a bounded rating scale. From the quantitative judgments it is possible to derive cross-modality matching functions, which specify physical intensity levels on the two modalities whose sensations are judged as equal. Stevens and Marks reported magnitude matches obtained in a sample of a dozen subjects. In order to match a light whose luminance was about 3 cd/m^2, the average matching SPL ran from 65 to 73 dB, depending on the experimental condition (as discussed below), with standard deviations averaging about 5 dB (the range of matching values averaged 16 dB). It would not be unusual, therefore, to discover individuals whose absolute cross-modality matches differ by 10 dB or more.

Perhaps even more significantly, Stevens and Marks discovered that the subjects' matches were reasonably consistent with repeated testing in different experimental conditions. In the three conditions, the range of light intensities remained constant, while in one condition the range of sound intensities was large, in another it was small and biased toward low SPLs, and in the third it was small and biased toward high SPLs. Across the different conditions, subjects revealed a modest degree of reliability, especially between the results obtained with the two small ranges of SPL (the correlation coefficient $r=0.66$). So the variation among people does not simply represent "noise" in the psychophysical measurements, but instead reveals stable individual characteristics.

I confirmed this finding in a more recent study (Marks, unpublished data), which compared the perception of intensity in hearing (using bands of low-frequency noise) and in taste (using solutions of sodium chloride). Again, the method was magnitude matching – the subjects gave numerical estimates of loudness and of taste intensity. And again all of the subjects served in multiple sessions – in this case, the sessions numbered four, with identical SPLs in all sessions, but with two different ranges of concentration of sodium chloride. In the four successive test sessions, each subject either compared the constant set of noises to concentrations of salt that were low, then high, then low again, and finally high again; or compared the set of noises to concentrations that were first high, then low, then high, and finally low. By using this paradigm I was able to examine two effects at once. First, what is the effect on magnitude matches of changing stimulus context? In other words, how, if at all, do magnitude matches change when the levels of salt concentration change from low to high or from high to low? And second, what is the effect of mere repetition of the same stimulus levels? In other words, how repeatable are a given person's magnitude matches when the same low or high levels are judged on two different occasions?

Two findings emerged. First, context did exert a potent effect (as I have observed elsewhere: see Marks, Szczesiul, & Ohlott, 1986; Marks, Stevens, Bartoshuk, Gent, Rifkin, & Stone, 1987). The intensity levels that subjects judged to match did change markedly when the set of stimulus intensities shifted. But more pertinent to the present argument, I found evidence of reliable interindividual differences; with repeated testing using identical sets of stimuli, people tended to be consistent in matching weaker or stronger solutions of sodium chloride to a given sound intensity (on average, $r = 0.62$). The statistical evidence confirms some casual observations and conversations I have had with subjects, to wit, that certain people will complete an experimental session by declaring that the most intense taste sensations they judged were far greater in magnitude than the loudest sounds, but with others declaring just the opposite, that the sounds were by far greater in magnitude than the tastes. As my colleague Linda Bartoshuk for a long time has argued, each of us lives in at least a somewhat unique perceptual world.

Individual differences become even more intriguing, in my view, when the sensory experiences being compared are multidimensional. In synesthesia, you may recall, and in cross-modality matching, it turns out that both increasing pitch and increasing loudness correlate with increasing visual brightness. Dim lights are low-pitched where bright lights are high; moreover, dim lights are soft where bright lights are loud. What happens, though, when subjects are given the opportunity to combine pitch and loudness in matching to brightness?

I attempted to answer this question in a couple of different ways (Marks, in press). In the first of two experiments, I displayed two lights, one dim and one bright, while I presented to the subjects pure tones of varied frequency and intensity; on each trial, the subject's job was to indicate which of the two lights more closely resembled a tone. The results revealed both the universal principle underlying cross-modal resemblance and the individual variations permissible within the general scheme. In general – but very much in general – the proportion of "bright" responses increased both with increasing sound frequency and with increasing sound intensity. In other words, higher pitch and higher loundess are both analogues to greater brightness. Nevertheless, people can and do differ substantially with regard to the degree that they rely on the one auditory attribute or the other. Some people are "pitch matchers," while others are "loudness matchers." Of the 16 participants in the experiment, half based their cross-modal judgments almost exclusively on pitch, giving greater proportions of "bright" response to higher sound frequencies, but always doing this independently of sound intensity; a couple of them based their judgments mainly on loudness, giving greater proportions of "bright" response to higher sound intensities, doing this largely independently of sound frequency; and the remainder of the participants used pitch and loudness about equally, giving greater proportions of "bright" response both to the higher frequencies and to the higher intensities.

Figure 1 shows typical examples, presented in the form of "equal-brightness curves." In order to derive these curves, I computed, for each subject, the various combinations of sound frequency and sound intensity that were judged equally similar to the dim and the bright lights; thus each data point indicates a single frequency-intensity combination that falls on a contour of "constant brightness." The interindividual differences among these contours are striking. For subjects 1 and 2 the curves are flat, representing constant sound frequency, and meaning thereby that only pitch mattered. Thus for subjects 1 and 2, when brightness is constant, pitch is constant. For subject 16 the curve is nearly vertical, representing nearly constant sound intensity, and meaning thereby that mostly loudness mattered. Thus for subject 16, when brightness is constant, loudness is roughly constant. And for subject 12, the equal-brightness curve is intermediate in form, having a negative slope, and implying thereby that both pitch and loudness mattered. Thus for subject 12, when brightness is constant, neither pitch nor loudness is constant.

It may seem surprising to find more people emphasizing pitch than loudness in matching sounds to lights, given the great extent to which psychophysical research and theory alike focus on the dimension of "intensity." But I will note, without additional comment, that synesthetic perceivers reveal exactly the same preference: pitch-brightness resemblance is the sine qua non of visual-auditory synesthesia.

The point I want to make is not just that individuals vary in the extent to which they rely on pitch or loudness in cross-modality matching; more important is the observation that they do so consistently. This consistency revealed itself in two ways. First, consistency was evident in the matches themselves, in that every subject served in four sessions testing different conditions, each condition using a different combination of sound intensities and frequencies. By and large, every subject tended to use pitch, loudness, or both in a con-

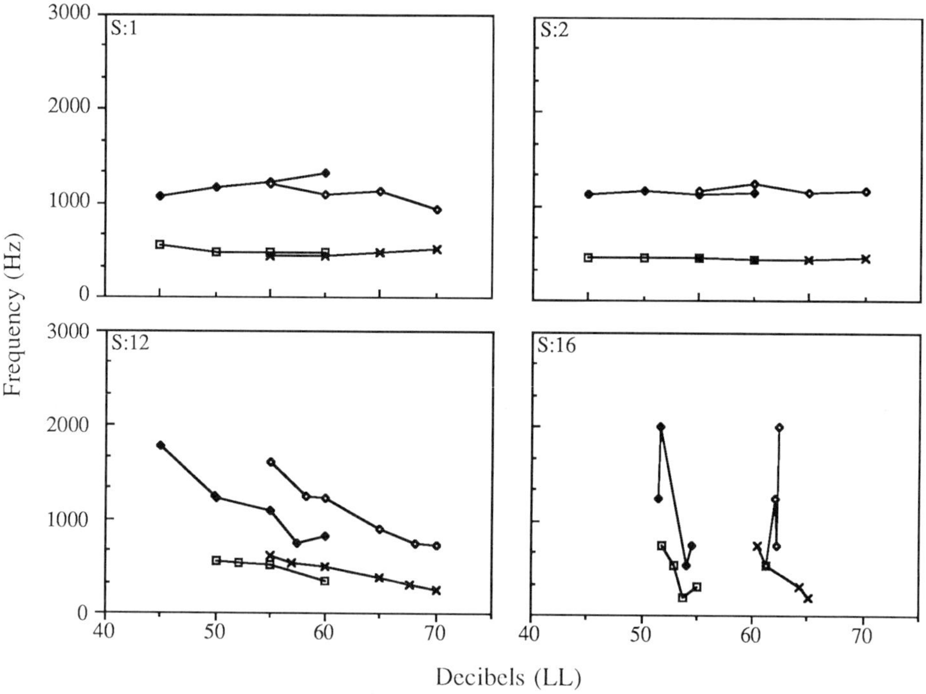

Fig. 1. Lines of constant cross-modal similarity between tone and light. Each *point* shows a tone that was judged equally similar to a fixed dim light and a fixed bright light, as determined from interpolations on linear psychometric functions. For subjects 1 and 2, the fact that each line of constant similarity is horizontal indicates that similarity was constant when pitch was constant, independent of loudness. For subject 16, the near-vertical curves indicate that similarity was constant when loudness was near constant, largely independent of pitch. For subject 12, loudness and pitch combined their effects in matching to brightness

sistent fashion over all four conditions; this consistency is evident in Figure 1, where the four symbols represent results obtained in the four conditions. Second, the consistent individual differences transcended the particular choice of psychophysical task, as I shall now describe.

Eight of the same subjects who served in the matching experiment also served in a study of cross-modality similarity scaling (Marks, in press). For this purpose, I used a set of 14 stimuli, nine pure tones differing in frequency and intensity and five white lights differing in luminance. Each trial posed for the subject two stimuli, sometimes two lights, sometimes two tones, and sometimes a light and a tone. The subject's task was to rate the degree of similarity between the two stimuli on a graphic scale, a line anchored by the labels "extremely dissimilar" and "extremely similar." Once I had a matrix of similarity judgments given to all possible pairs of stimuli, it was possible to submit the results to multidimensional scaling in order to determine the best spatial arrangement of the stimuli consistent with the judgments; that is, to determine the best arrangement in Euclidean space of points whose interstimulus distances matched the original judgments of dissimilarity. To this end I adopted the INDSCAL model of Carroll and Chang (1970), which has two elegant features. For one, INDSCAL provides measures of individual differences in the weightings of the component perceptual dimensions, and for another, it uses these individual differences in weighting coefficients to identify the orientation of the perceptual axes in the multidimensional space.

Multidimensional scaling (using the computer program SINDSCAL) showed the similarity judgments to be consistent with the two-dimensional representation depicted in Fig. 2. As the figure indicates, the dimension on the abscissa corresponds to loudness (the tones increase in intensity from left to right) and the dimension on the ordinate corresponds to pitch (the tones increase in frequency from bottom to top). The lights fall within the same two-dimensional space, forming an arc of increasing brightness that runs from the lower left to the upper right, and thus cutting through both the loudness axis and the pitch axis. Increasing brightness corresponds both to increasing loudness and to increasing pitch. So it is perhaps more appropriate to identify the dimension on the abscissa as representing *loudness–brightness* and the dimension on the ordinate as representing *pitch–brightness*. Note a curious and striking corollary to these results – brightness does not require its own perceptual dimension within the representational space. That is, the psychological structure of the similarity judgments does not separate the stimuli by modalities. Sounds and lights alike fall within a common, bimodal psychological representation.

The spatial map shown in Fig. 2 gives the configuration determined for the group of eight subjects as a whole. But the scaling results also provide information about individual differences among the subjects, and in particular, information about differences in the magnitudes of the weighting coefficients. On average, given the particular stimuli I used, the weighting coefficients were about the same on the two dimensions; but individuals varied markedly. Some subjects gave much greater weighting to the dimension of loudness–brightness, while others gave much greater weighting to the dimension pitch–bright-

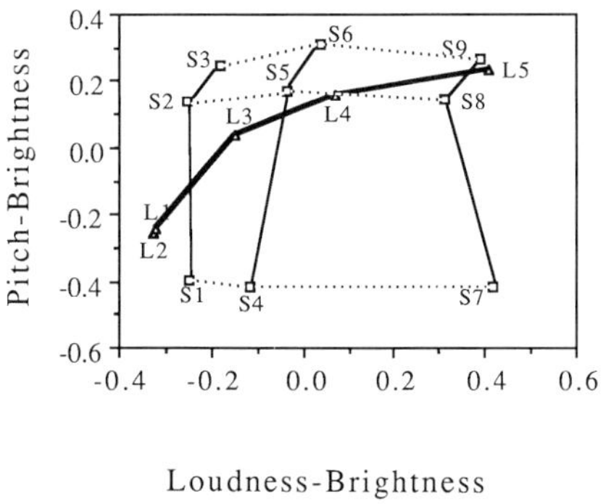

Fig. 2. The underlying multidimensional representation of perceptual experiences of lights (*vision*) and tones (*hearing*). Dim lights resemble low-pitched and soft tones; bright lights resemble high-pitched and loud tones. *L1–L5* indicate lights of increasing brightness. *Continuous lines* connect sounds (*S*) of equal loudness, loudness increasing from *left* to *right; dotted lines* connect sounds (*S*) of equal pitch, pitch increasing from *bottom* to *top*. Individuals vary in the extent to which their personal representations emphasize one dimension or the other

ness. To establish the multidimensional space for a given individual, then, one axis or the other in Fig. 2 must be stretched or compressed. Finally, and most pertinent to the general argument, the individual weighting coefficients determined from multidimensional scaling correlated highly with measures determined on the same subjects in the cross-modality matching experiment (the rank-order correlation coefficient ϱ equaled 0.90). In other words, the salience of auditory pitch or loudness as a cross-modal analogue to visual brightness characterizes each individual, transcending the particular psychophysical method used to study similarity.

Through the main portion of this article, I have recounted a few of the general rules and the individual variants that characterize the perception of cross-modal similarity. My own research has produced a few surprises, even with respect to the general rules. For roughly a century now, since at least the time of Külpe (1893), psychophysicists have pointed to the significance of perceived intensity as a dimension common to all sensory modalities. But in the realm of visual-auditory similarity, the perceived intensity of light – the dimension of brightness – does *not* necessarily find its closest or strongest analogue in the perceived intensity of sounds – the dimension of loudness. Instead, for many of us, maybe most, brightness finds its nearest relative in pitch. Both pitch and loudness contribute to cross-modal similarity, for similarity is fundamentally multidimensional. And within the multidimensional framework established by general rules of psychophysical correspondence, there is ample room for individual differences to flourish.

But is it so surprising to learn that individuals can vary widely or strikingly in their perception of cross-modal similarity, despite the presence of an underlying unity? After all, the notion of similarity itself both entails identity and at the same time denies it – we say that A *resembles* B to the degree that or in the way that they are the same, but we say that A *only resembles* B in that they differ. Just in this fashion does a general principle both entail a communality among its individuals yet acknowledge the singularity of each member.

The sum total of these studies tells a tale of a type we have heard before, a tale of universals and particulars, in this way resembling many other dialectical stories, some philosophical, some scientific, and some literary. Like other branches of natural science, sensory psychophysics reveals general principles that come in a variety of flavors. In doing this, psychophysics provides a conceptual framework to satisfy the most singleminded hedgehog among us: the framework of psychophysics serves to establish those synchronic rules that describe uniformities in perceptual behavior, uniformities that emerge under wide ranges of conditions and that seemingly apply to most if not all people. But at the same time sensory psychophysics reveals a universe as richly diverse as each of our personal set of perceptual experiences. So psychophysics can nourish the most eclectic fox among us. For psychophysics makes its pallet under the roof of psychology, that quintessential human science that ultimately must pay homage to each person's diachronic heritage, to each person's essential uniqueness.

References

Berlin, I. (1953). *The hedgehog and the fox.* New York: Simon and Schuster.
Bleuler, E., & Lehmann, K. (1881). *Zwangsmässige Lichtempfindungen durch Schall und verwandte Erscheinungen.* Leipzig: Fues.
Borg, G. (1961). Interindividual scaling and perception of muscular force. *Kungliga Fysiografiska Sällskapets: Lund Förhandlingar 31,* 117–135.
Borg, G. (1962). *Physical performance and perceived exertion.* Lund, Sweden: Gleerup.
Borg, G. (1972). A ratio scaling method for interindividual comparisons (Report No. 27). University of Stockholm, Institute of Applied Psychology.
Carroll, J. D., & Chang, J. J. (1970). Analysis of individual differences in multidimensional scaling via an *n*-way generalization of "Eckhard-Young" decomposition. *Psychometrika 35,* 283–315.
Flournoy, T. (1893). *Des phénomènes de synopsie.* Paris: Alcan.
Külpe, O. (1893). *Grundriss der Psychologie.* Leipzig: Engelmann.
Lewkowicz, D. & Turkewitz, G. (1980). Cross-modal equivalence in early infancy: Auditory-visual intensity matching. *Developmental Psychology, 16,* 597–607.
Marks, L. E. (1974). On associations of light and sound: the mediation of brightness, pitch, and loudness. *American Journal of Psychology, 87,* 173–188.
Marks, L. E. (1975). On colored-hearing synesthesia: cross-modal translations of sensory dimensions. *Psychological Bulletin, 82,* 303–331.
Marks, L. E. (1978). *The unity of the senses: Interrelations among the modalities.* New York: Academic.
Marks, L. E. (in press) On cross-modal similarity: the perceptual structure of pitch, loudness, and brightness. *Journal of Experimental Psychology, Human Perception and Performance.*
Marks, L. E., Borg, G., & Ljunggren, G. (1982). Individual differences in perceived exertion assessed by two new methods. *Perception and Psychophysics, 34,* 280–288.
Marks, L. E., Szczesiul, R., & Ohlott, P. J. (1986). On the cross-modal perception of intensity. *Journal of Experimental Psychology: Human Perception and Performance, 12,* 517–534.
Marks, L. E., Hammeal, R. J., & Bornstein, M. H. (1987). Perceiving similarity and comprehending metaphor. *Monographs of the Society for Research in Child Development,* Vol. 52 (Serial No. 215).
Marks, L. E., Stevens, J. C., Bartoshuk, L. M., Gent, J. G., Rifkin, B., & Stone, V. K. (1987). Magnitude matching: the measurement of taste and smell. *Chemical Senses, 13,* 63–87.
Stevens, J. C., & Marks, L. E. (1980). Cross-modality matching functions generated by the method of magnitude estimation. *Perception and Psychophysics, 27,* 379–389.

Deceptive Psychophysics: The Probe-Signal Method and Focused Attention *

Bertram Scharf, Camille-Aimé Possamaï, and Anne-Marie Bonnel

Abstract

Psychophysicists nearly always seek to have the observer pay full attention to the stimuli that he is to detect or discriminate. What about performance with stimuli that receive less than full attention or that may even be wholly unexpected? The probe-signal method and similar procedures permit the measurement of performance outside the focus of attention. In a typical auditory paradigm, the observer is to detect a tone at a designated frequency. On most trials the signal is presented at that frequency, but on some trials, say 25%, the signal is presented at a different frequency. Many data show that the observer detects the attended and nearby frequencies very well but misses frequencies more than about half a critical band away. For example, detection of an expected tone at 1000 Hz is 90% and at 1500 Hz is 55%. Most of the data are readily explained by a single-band model. However, when auditory space is examined, signals are found to be detected just as well from unexpected as from expected directions. Applied to vision, the probe-signal method reveals an advantage for an expected spatial frequency, spatial position, or line orientation. Other related approaches have shown strong effects of selective attention in the discrimination of consonants (*da* and *ga*) but little effect on loudness estimation. The methods permit quantitative measurements of cognitive influences on psychophysical outcomes; and despite the title, they do not require deception.

Introduction

This paper is concerned with how attention affects detection and discrimination in hearing and seeing. Normally, psychophysicists strive to have their observers pay full attention to a particular stimulus presented at a specified time and place. When attention is the variable of interest, psychophysicists and cognitive psychologists seek to manipulate the distribution of attention across stimuli. One approach, exemplified by the probe-signal method of Greenberg

* Preparation of this paper was in part supported by grants NS07270 from NIH and 86/649 from the Scientific Affairs Division of NATO.

and Larkin (1968), is to focus the observer's attention on one "primary" stimulus (defined by type, time, or place) and then measure performance with some other "probe" stimulus. Except for the probes, the probe-signal experiment resembles a typical psychophysical experiment. The observer is enjoined to focus all his attention on a well-defined signal, so that the distribution of attention is 100:0, with all attention directed to the primary and none to the probe. The beauty of this method is that it maintains the 100:0 distribution over many trials in which the probe signal is presented. Originally, it seemed important to keep observers ignorant about the existence of probes – hence the title of this paper –, but deception turns out to be unnecessary.

Other approaches seek less extreme distributions of attention, such as 80:20 or 50:50. One paradigm popular in vision has each trial begin with an indication of the probable locus of the signal, for example, whether to the left or right of fixation. Observers generally react more quickly to a signal on the more probable side (Posner, Nissen, & Ogden, 1978), to which they seem to pay more attention. One investigation (Bashinski & Bacharach, 1980) seemed to indicate that also threshold is lower on the more probable side, but a recent investigation (Müller & Findlay, 1987) casts doubt on that interpretation. In another approach, the dual-task paradigm, attention is divided between two simultaneous or concurrent tasks (Sperling, 1984). Instructing their observers to pay different proportions of attention to each task, Bonnel, Possamaï, and Schmitt (1987) showed that discrimination is better on the task that receives more attention; on the other hand, detection is unaffected by the allocation of attention (Bonnel, Possamaï, & Scharf, 1987).

We confine this review mostly to focused attention as achieved by the probe-signal method and as applied to detection. Also discussed are studies that measured reaction time, speech discrimination, and loudness. (We hope the loudness study will please Gunnar Borg who contributes so much to the measurement of sensation magnitude.) After describing the probe-signal method and reviewing auditory studies that have used it and similar paradigms, we turn to analogous studies in vision.

Auditory Studies

Nearly all the psychoacoustical studies employing the probe-signal method have varied the frequency of a pure-tone signal. The first section reviews those studies. The second section reviews studies that varied temporal and spatial locus. The third section describes experiments on loudness scaling and speech perception.

Frequency

The probe-signal method appears to have been inspired by a report by Tanner and Norman (1954; cited by Swets & Kristofferson, 1970). Using a four-interval forced-choice test, they set the level of a 1000-Hz tone burst presented in white noise so that their four unpracticed observers scored 65% correct. After a break, they changed the signal frequency to 1300 Hz, without informing the observers. Percentage correct dropped to chance, 25%. Telling the observers of the frequency change brought performance back to 65%. The implication was that the observers, listening for a 1000-Hz tone, had missed the 1300-Hz signal. It was not until 1968, however, that Greenberg and Larkin published an explicit test of that implication, based on their probe-signal method.

Greenberg and Larkin (1968) trained their observers to listen for a 250-ms, 1000-Hz tone burst in a wide-band noise. They kept the observers focused on the primary at 1000 Hz, usually by presenting it at a level well above threshold at the beginning of a block of trials. The signal was then set at a level that yielded approximately 80% correct (two-interval forced choice). On 23% of the trials, a probe tone burst at a frequency other than 1000 Hz was presented, without informing the observer; the frequency of the probe was the same throughout a block of 75–100 trials. Means for seven observers are shown as the dashed curve in Fig. 1c. Results were unambiguous. Scores were close to 80% correct when the signal was at or near the expected frequency, and then fell rapidly at frequencies farther away. The observers seem to have listened

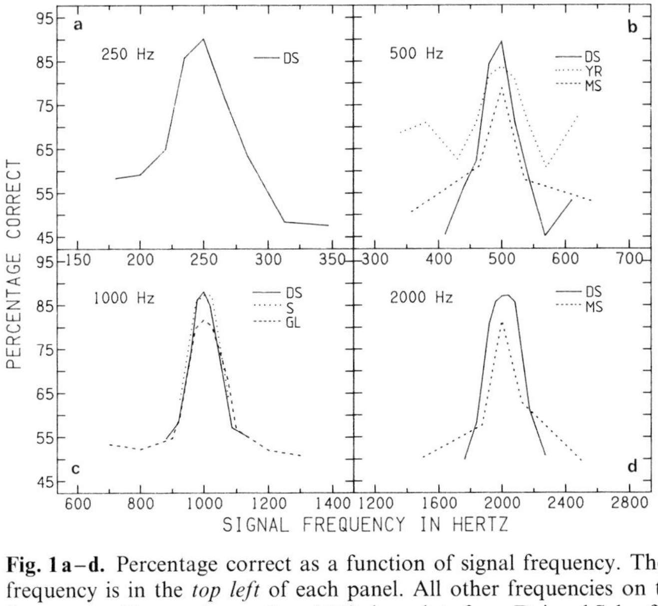

Fig. 1a–d. Percentage correct as a function of signal frequency. The primary or expected frequency is in the *top left* of each panel. All other frequencies on the abscissa are probe frequencies. The *continuous lines* (*DS*) show data from Dai and Scharf (1988). The other data come from: *GL*, Greenberg & Larkin (1968); *MS*, Macmillan & Schwartz (1975); *S*, Scharf, Quigley, Aoki, Peachey, & Reeves (1987); *YR*, Yama & Robinson (1982). At all four primary frequencies and in every study, primary and nearby probes are detected well and far probes are detected poorly if at all. Moreover, the relative width of the filter-like functions is roughly constant and equal to the critical band

within a band of frequencies, centered on the primary, which was roughly the size of the critical band (165 Hz at a center frequency of 1000 Hz; Scharf, 1970). The most plausible inference is that listeners who focus on a particular frequency are less sensitive to signals outside the surrounding critical band than to those inside.

Yama and Robinson (1982) ran a nearly identical experiment with the primary signal set at 500 Hz. Probe frequencies were between 340 and 620 Hz. Even after many trials with each of the probes, observers were surprised to learn that frequencies other than 500 Hz had been presented. Means for the six well-trained observers are shown as the dotted curve in Fig. 1 b. Once again, bandwidth is similar to that of the critical band, which is approximately 100 Hz at a center frequency of 500 Hz.

Scharf, Quigley, Aoki, Peachey, and Reeves (1987) also used the probe-signal method. Unlike the few, highly trained observers of the two earlier studies, Scharf et al. tested many unpracticed observers, each of whom served in a single session. To focus the observer's attention on the 350-ms, 1000-Hz primary, the experimental session began with a block of 100 trials in which only the primary was presented, at 4 dB above threshold in a wide-band noise at 60 dB. Moreover, at the onset of each trial, the 1000-Hz tone was presented as a cue, 8 dB above threshold. In the next two blocks, the 1000-Hz cue was presented on every trial, the 1000-Hz signal was presented on 75 trials, and a probe at a different frequency was presented on 25 trials. Results were like those of Greenberg and Larkin (1968) with detection near 90% in the vicinity of the primary and near 50% well outside the surrounding critical band, at 600 and 1500 Hz. The dotted curve in Fig. 1 c shows some of the data from Scharf et al. (1987).

In a modified version of the probe-signal method, Dai and Scharf (1988) present signals at eight probe frequencies in a block of 64 trials. Each probe is presented only twice in a block while the primary is presented 48 times. A warning cue at the primary frequency initiates each trial. The mean results of three to four, well-trained observers are presented as the continuous curves in each of the panels of Fig. 1. Data are for 120 to almost 400 judgments per probe frequency. Results are like those with a single probe in a block.

Macmillan and Schwartz (1975) also collected data by the probe-signal method at various frequencies. Their results were transposed down from primaries at 700 and 2300 Hz and are plotted as the dashed curves in Fig. 1 b, d. Based on few points, tuning appears somewhat sharper, but is essentially the same as in the other studies.

Figure 1 reveals striking agreement among various studies. Results differ little whether observers are trained or untrained, whether cues occur on every trial or not, whether a single probe or many probes come in a block of trials, and whether observers are informed or not. (Two of the four observers of Dai and Scharf were fully informed and the other two were wholly naive; yet no consistent pattern of differences emerged; see also Scharf et al. 1987.) Of course, important individual differences do exist and remain to be explored.

Nevertheless, the average results of Fig. 1 reveal the general effect of focusing attention on a particular frequency region.

The two-interval forced-choice procedure was common to the above experiments. A single-interval yes–no procedure yields similar results (Scharf, Canévet, Possamaï, & Bonnel, 1986; Scharf, 1988). With the primary at 1000 Hz, sensitivity as measured by the index d' of signal detection theory was 0.6 for a distant probe at 1500 Hz, down from 2.0 when the same tone was the only signal in the control condition. A near probe at 1025 Hz showed an insignificant decrement, from 2.2 to 2.0.

These data support a single-band model of selective auditory attention. Listeners seem to focus their attention in a narrow band of frequencies roughly the width of the critical band and to detect signals within that band better than signals outside it. The model easily accommodates the possiblity that listeners recenter their listening band according to experimental demands or vary the width somewhat. However, some results also suggest that listeners can use more than a single band. Macmillan and Schwartz (1975) used the classical single-probe method (two-interval forced choice) with *two* simultaneous primaries. They trained their observers to listen for primaries far apart at 1000 and 2300 Hz or, in a separate series, at 700 and 1600 Hz. While both primaries were detected on 80% of the trials, performance was at chance for far probes, even for those lying *between* the two primaries. The widths of the listening bands were roughly those of the relevant critical bands. That observers did not simply switch from one primary frequency to the other is shown by fact that the percentage correct on the two primaries was nearly as high whether observers had to listen to two primaries or to one.

The results of several other studies that did not use the probe-signal method are also most readily interpreted as showing that observers can listen to more than a single band at a time (Buus, Schorer, Florentine, & Zwicker, 1986; Johnson & Hafter, 1980). The number of monitored bands depended in these studies on the number of signals or the presence of multiple cues. Penner (1972) showed that payoffs could also alter the number of attended bands. Payoffs were set so that observers would benefit maximally by listening in one condition primarily to a single frequency and in another condition to many frequencies. All signal frequencies were equally probable in all conditions. Detection was best at the target (and cue) frequency when payoff was maximized at that frequency; it was approximately the same at all frequencies when payoff was maximized for detecting all frequencies equally well. These results lend further support to the notion that listeners can adjust their listening strategy to fit experimental demands. It is unclear just what new strategy is adopted, whether a single band that moves rapidly from one frequency region to another, a wider listening band, multiple bands, or some combination of these. And individual observers seem to adopt different strategies (Scharf et al. 1987; Sorkin, Pastore & Gilliom, 1968).

Space and Time

Listeners can be led to listen for a sound from a particular direction or at a precise point in time. Only a single study appears to have used the probe-signal method to measure possible effects on the detection of sounds when they come from an unexpected direction, and another study when they come at an unexpected moment.

Scharf et al. (1986) had their observers detect a tone burst that was presented 4 dB above its masked threshold. The signal came from a loudspeaker 45° to the observer's left in a first block of 100 trials. Each trial comprised a warning cue, which was the same signal at 8 dB from the left, and two observation intervals indicated by lights. The observer reported which interval contained the signal. In the second block, the primary was a signal from the left and occurred on 80% of the trials and the probe was the signal from a loudspeaker 45° to the right on 20% of the trials. In the third block, the cue and all signals came from the right. Six observers scored 80%–90% correct for both primaries and probes in all three blocks. Thus, detection of the probe was just as good whether it came from an unexpected or from an expected (and presumably attended) direction. (However, observers do respond more rapidly to sounds from an expected direction when they must make a spatial or frequency discrimination; Rhodes, 1987; Scharf et al. 1986, 1988).

Leis-Rossio (1986) used the probe-signal method to measure the detection of signals presented at unexpected times, while the subject is focusing attention on an expected or primary temporal locus. The task was to detect a 0.5-ms rectangular pulse presented monaurally in a continuous wide-band noise. A single-interval yes–no procedure was employed. Listening at the primary locus – 500 ms after onset of the observation interval – was fostered by presenting ten times in a row the signal well above threshold at that primary locus. In the experimental condition, most of the signals were presented at the primary locus; the rest were presented at one of six probe loci from 100 to 1300 ms. Figure 2 gives mean results for five observers. The data seem to provide evidence for a temporal window or filter that sets the temporal limits for the observer's attention or listening band. However, there remains the thorny problem as to whether the observers' criteria may have been most lax at the expected temporal location where signal probability was highest.

Loudness and Speech Perception

The distinguishing characteristic of the probe-signal and similar methods is that they seek to focus a listener's attention on one stimulus and then measure performance on other stimuli as well. We now consider two performance measures other than detection, loudness and syllable discrimination.

Ullman and Scharf (1975) had observers play a television ping-pong game against the experimenter while a series of noise bursts came from a loudspeaker every 2 s. Their levels ranged in 10-dB steps from 20 to 80 dB SPL. The observer's auditory task was to estimate the loudness of the noise burst that

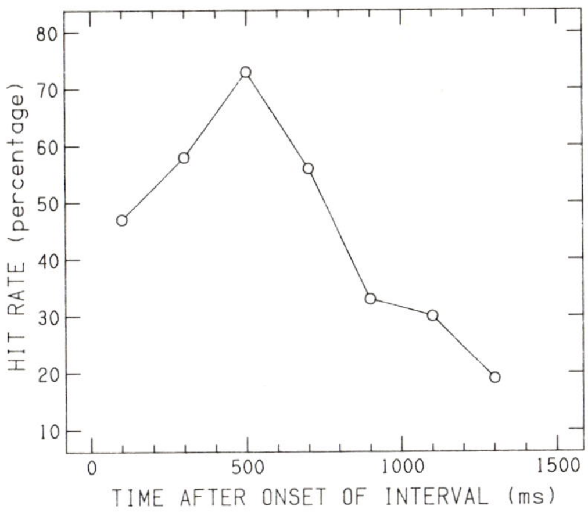

Fig. 2. Hit rate (percentage correct on signal trials in a yes–no paradigm) plotted as a function of the time at which the signal was presented after onset of the observation interval. Data are the means of five observers. Observers expected a signal at 500 ms, the temporal locus at which the hit rate was greatest. (With permission, from Leis-Rossio, 1986.)

was on just prior to a darkening of the television screen, which interrupted the game. Judgments were called for at irregular intervals – with no warning to the observer – varying from 20 to 44 s. On catch trials, no sound was presented just prior to the darkening of the screen. In a separate session observers estimated the loudness of noise bursts that came on just *after* the screen had darkened. It was assumed that subjects paid full attention to the sounds in the latter, attention condition and little if any in the former, nonattention condition. Thus, all the signals were "probes" in the nonattention condition. On the whole, estimated loudness was the same power function of sound pressure in the two conditions; the functions differed only at the two lowest levels of 20 and 30 dB where signals were judged on the average equal in loudness or louder when not attended than when attended. Our tentative suggestion is that listeners absorbed by a visual task often do not hear weak sounds and judge earlier, louder sounds in their place. They do hear sounds at higher levels (40–80 dB SPL) with their usual loudness.

In the detection experiments above, observers were led by examples or repeated cues to focus on a particular sound. In experiments on speech perception, a natural and automatic focusing of attention takes place. Presented with speech signals, observers listen in a phonetic mode; they attend to phonetic cues (e.g., Best, Morrongiello, & Robson, 1981; Repp 1982, 1985). However, various manipulations can lead observers to listen in an auditory mode, and attend to nonspeech cues. Santon and Scharf (1987) report a particularly revealing example of these differences when French and American listeners had to discriminate synthesized *da*s from *ga*s. The syllables differed only in that for *da*, the frequency of the initial noise burst and the onset frequency of the third-formant transitions were located near 3000 Hz, whereas for *ga*, they were near 2000 Hz. After 50 ms, the third formants of both syllables reached the vocalic steady-state value of 2500 Hz. Discrimination was measured in the presence of narrow-band masking noise set at center frequencies from 1000 to

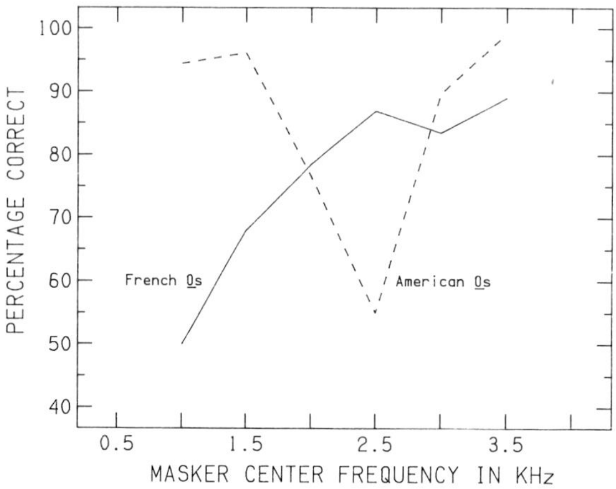

Fig. 3. Percentage correct discrimination between synthesized *da*s and *ga*s as a function of the center frequency of a 300-Hz wide band of masking noise. Data are means for three observers whose native language was American (*dashed line*) and two observers whose native language was French (*continuous line*)

3500 Hz. For three well-trained American observers, percentage correct was near chance when the noise was centered on 2500 Hz and rose rapidly for noises at lower and higher frequencies. Their data describe the dashed function in Fig. 3.

Results were very different for two French listeners, as shown by the continuous line. Percentage correct was at chance when the noise was centered on 1000 Hz and rose steadily as the noise moved to higher frequencies. Moreover, the two French observers were more sensitive and required a much lower speech-to-noise ratio than did the Americans. The interpretation of these results is based on preliminary spectral analyses, the effect of feedback, and introspective reports. The French appear to have heard the *da*s and *ga*s as poor speech sounds. Accordingly, they listened in the auditory mode and picked up a nonphonetic cue, apparently a low-frequency artifactual click that accompanied only the *da*s. Being in a phonetic mode, the American observers focused attention on the 2500-Hz region where all the phonetic information lay and therefore missed the "probe" at the lower frequencies. A careful test is needed to confirm the hypothesis that listening in a phonetic mode can lead observers to miss nonphonetic signals in the same way that they miss probes in the standard probe-signal experiment.

Visual Studies

The probe-signal method has been used only occasionally in vision, probably because deception, which had been thought indispensable, is more difficult in vision than in audition, at least for simple stimuli presented centrally. For example, whereas the observer is unaware of a weak 1500-Hz tone when expecting a 1000-Hz tone, it seems unlikely that he would miss a red patch when expecting a green one. Nevertheless, a few vision studies have used the probe-signal method or one like it to study spatial frequency, motion perception, flash detection, and detection of line segments.

The theoretical questions that these experiments address are similar to those posed in audition. They include questions about tuning or filter characteristics and about whether a single-band or multiband model better accounts for the data. The question also arises as to whether observers can switch rapidly (within a trial) from one filter to another. Much evidence suggests the existence of selective mechanisms or filters devoted to the analysis of the various features of visual stimuli. For example, one set of filters is devoted to the analysis of spatial frequencies, another to direction of movement, and so on. Each filter is maximally sensitive to a given value of the stimulus, decreasing monotonically, and probably symmetrically, on both sides of the maximum.

Given these filters, it remains unclear whether the observer can use several of them at the same time, or is limited to the use of one at a time. In other words, is the observer able to monitor simultaneously several possible values of the stimulus (multiband model) or only one (single-band model)? If only a single band can be monitored, can the observer then switch rapidly from one band to another?

Spatial Frequencies

Davis and Graham (1981) used the probe-signal (two-interval, forced choice) procedure to investigate different attention strategies in the detection of sinusoidal gratings. They compared d' when each of four or seven different frequencies were presented alone in each of a block of 120 trials and when they were intermixed. The intermixed blocks began with the presentation of a primary frequency at a high contrast; observers were encouraged to attend that frequency on the rest of the trials. In three conditions, the signal was at the primary frequency on 80% of the trials, and at one of six other "secondary" frequencies on the other 20%. In a fourth condition, the primary occurred on 95% of the trials, and one of three secondary frequencies on the rest. Detection was generally better, relative to what it was in the "alone" blocks, at the primary frequency and decreased with increasing separation between the probe and primary frequencies. The authors interpreted their results as supporting a single-band model in which observers monitored a single frequency band when instructed to attend to a particular frequency. Sequential analyses of the data suggested further that in the 80% conditions, observers slowly switched their listening band. On the whole, the tuning elicited for visual spatial frequencies appeared much less sharp than for auditory signal frequencies. The advantage for the primary seldom exceeded 20% or 25% in the visual studies, whereas it was often close to 40% in the auditory studies. Although this difference may reflect higher performance levels in the hearing studies at the primary frequencies, we suspect that probes at distant spatial frequencies presented in the same location as the primary are more difficult to tune out than are probes at distant auditory frequencies.

Direction of Movement

Ball and Sekuler (1980, experiment 2) used a kind of probe-signal procedure to measure reaction time to the onset of movement. The stimulus was a large array of random dots on a cathode-ray tube. All the dots began to move in the same direction after a foreperiod lasting 2–3.5 s. The observer responded as quickly as possible when he saw the dots begin to move; he did not report the direction of movement. Three conditions were run: (1) the direction was the same throughout a block of 50 trials, either 60° or 120°; (2) direction was one of two on every trial, 60° or 120°; (3) direction was one of three on every trial, 60° or 120° on 90%–95% of the trials, 90° on 5%–10%. The observer always knew in every condition about the stimulus probabilities. In effect, then, the third condition had two primaries and a probe at 90°. The motivation for this configuration was an earlier experiment that suggested that observers focus on a direction midway between two expected directions. Results fully supported the midway hypothesis. Reaction time was 6%–10% longer to the two primary directions than to the infrequent probe direction. This outcome is quite different from that of Macmillan and Schwartz (1975) who found that observers monitored two distant auditory frequencies and missed those lying between. The basis for the different outcomes is unclear, whether modality, observers, procedures, or performance measure (reaction time vs detection).

Spatial Position

Several studies have shown that observers can orient attention toward a signal in the peripheral visual field, without moving their eyes (for a review see Possamaï, 1986). These shifts of attention may be likened to an attentional beam moving across the visual field. Two main characteristics of the beam have been described. First, it cannot be split between noncontiguous regions (e.g., Posner & Cohen, 1984). The unity of the beam, however, holds only in single tasks; in a dual task, the observer can divide his attention between spatial locations (Bonnel, Possamaï & Schmitt, 1987). Second, the area where sensitivity is maximum covers no more than approximately 1° (e.g., Humphreys, 1981). Sensitivity decreases as the stimulus moves away from the center of the attentional beam, at least up to 5° of visual angle (Gatti & Egeth, 1978). Clearly, the attentional beam of vision is conceptually close to the critical band of audition. However, research on spatial effects in vision has been mostly unrelated to auditory research. Our own visual experiment took advantage of the probe-signal method and was designed to parallel auditory studies of the critical band. Moreover, we considered spatial position in vision as equivalent to frequency in audition.

The observer's task was to detect a luminance increment in a yes–no procedure. A fixation point was located at the center of a video camera lens used to monitor eye movements. To each side of fixation were two red, continuously energized light-emitting diodes, 5° and 7° to the left and 5° and 7° deg to the right along a horizontal row. In a given block, the same diode was brightly illuminated at the beginning of each trial. This diode served as the warning and to draw the observer's attention, without eye movement. The signal, a 10-ms weak luminance increment, came on 500 ms later but only on half the trials. The primary signal was an increment in the warning diode, and occurred on 75% of the signal trials. The probe was an increment in the diode on the other side of the fixation point (at the same distance from fixation), and occurred on 25% of the signal trials. Eight observers were tested under one condition with primary and probe 10° apart (near condition), and under another condition with them 14° apart (far condition). In a control condition, the signal was always in the warning diode; there were no probes. Besides responding yes or no, observers indicated their confidence on a 3-point scale for each response. Sensitivity was assessed using a nonparametric index, $P(A)$. Half the observers were told beforehand about the probes.

The results show that sensitivity to the near primary ($P(A)=0.93$), to the far primary (0.86), to the near probe (0.94), and in the control condition (0.91) did not differ statistically. Only sensitivity to the far probe (0.74) was significantly poorer. The absence of any effect in the near condition suggests a single band centered on the fixation point, with a maximum extent of between 10° and 14° of visual angle. Although the limit of the functional field (Ikeda & Takeuchi, 1975) seems to fall in this zone, the sharp difference we found between the near and far probes, which differed by only 2° is surprising. The effect may result from the configuration of the diodes rather than from their visual angles. While the near probes were contiguous, the far probes were separated by the near ones. These interposed diodes may have acted as added noise in the far condition. The visual display was then noisy enough to reveal an attention effect for the far probes, but not noisy enough for the near probes. Hence, the observed difference between the two conditions. This interpretation receives support from a recent experiment by Nougier (personal communication, 1987). He replicated our experiment with an additional 16 horizontal diodes, set 1.5° apart. The near probes were separated by 9°, the far probes by 15°. Sensitivity to both probes was significantly lowered and by the same amount. An obvious follow-up would be to shorten the distance separating the primary and probe to values comparable to those reported in other types of experiments on the attentional beam.

Line Orientation

In another standard application of the probe-signal method, Kurylo (personal communication) measured the detection of lines presented at two different orientations on a cathode ray tube. Twenty naive, untrained students served as paid observers in a single session. Each trial began with a line at a 45° orientation as the cue, followed by two intervals one of which contained a signal, which was the same line on 75% of the trials and a line at the opposite 45° orientation on the other 25%. The cue line was presented long enough to be clearly perceived, and the signal very briefly, just long enough so it could be detected most of the time. Whereas the primary was detected on 90% of the trials, the probe was detected on 81%. The advantage was in the same direction whether the primary had a negative or positive slope. While highly significant ($p < 0.01$), the difference of 8% is much smaller than reported in experiments on auditory frequency and visual spatial locus. The effect is small probably because the central portions of the primary and probe occupied the same position in space, approximately 0.25° above the fixation point. We should expect a larger effect with lines whose central portions are eliminated. (Of course, eye movements would have to be carefully monitored.)

Conclusions

In audition, the probe-signal method has provided a series of congruent results from several laboratories, all showing that focused attention to a particular frequency region reduces detection outside that region. Of special interest is that the width of the favored region is similar to that of the critical band as measured in a variety of psychoacoustical tasks. Attention focused on a particular temporal locus may also reduce detection of signals arriving earlier or later, but this conclusion remains tentative. On the other hand, focused attention does not affect detection of sounds coming from unexpected directions.

In vision, tuning under focused attention has been demonstrated for spatial frequency and suggested for spatial locus and line orientation. Many of the data support a single-band model of attention in both audition and vision. Nevertheless, observers seem to be able to switch to other strategies when instructions or payoffs make a different strategy more appropriate.

Despite the title of this article, deception is unnecessary to focus attention using the probe-signal method, at least when auditory frequency or visual locus is the stimulus variable. The method has proven extremely robust and reliable; its full potential remains to be explored.

Acknowledgements. We thank Huanping Dai and Josephine Todrank for producing the figures and Lori Jordan for preparing the manuscript on the word processor. We are also grateful to Boys Town Institute for providing the software for drafting the figures.

References

Ball, K. & Sekuler, R. (1980). Models of stimulus uncertainty in motion perception. *Psychological Review 87,* 435–469.

Bashinski, H.S. & Bacharach, V.R. (1980). Enhancement of perceptual sensitivity as the result of selectively attending to spatial locations. *Perception and Psychophysics, 28,* 241–248.

Best, C.T., Morrongiello, B. & Robson, R. (1981). Perceptual equivalence of acoustic cues in speech and non-speech perception. *Perception and Psychophysics, 29,* 191–211.

Bonnel, A-M., Possamaï, C-A., & Scharf B. (1987). *Spatial allocation of attention: line length discrimination vs. luminance detection.* Psychonomic Society, 6–9 November 1987. Seattle.

Bonnel, A-M., Possamaï, C-A., & Schmitt, M. (1987). Early modulation of visual input: A study of attentional strategies. *Quarterly Journal of Experimental Psychology, 39* (A), 757–776.

Buus, S, Schorer, E., Florentine, M., & Zwicker, E. (1986). Decision rules in detection of simple and complex tones. *Journal of the Acoustical Society of America, 80,* 1646–1657.

Dai, H. & Scharf, B. (1988). Frequency selectivity in focused attention as demonstrated by a multiprobe method. *Journal of the Acoustical Society of America, 83,* S 33.

Davis, E. T. & Graham, N. (1981). Spatial frequency uncertainty effects in the detection of sinusoidal gratings. *Vision Research, 21,* 705–712.

Gatti, S. W. & Egeth, H. E. (1978). Failure of spatial selectivity in vision. *Bulletin of the Psychonomic Society, 11,* 181–184.

Greenberg, G. & Larkin, W. (1968). Frequency-response characteristic of auditory observers detecting signals of a single frequency in noise: The probe-signal method. *Journal of the Acoustical Society of America, 44,* 1513–1523.

Humphreys, G. W. (1981). On varying the span of visual attention: Evidence for two modes of spatial attention. *Quarterly Journal of Experimental Psychology, 33* (A), 17–31.

Ikeda, M. & Takeuchi, T. (1975). Influence of foveal load on the functional visual field. *Perception and Psychophysics, 18,* 255–260.

Johnson, M. & Hafter, E. (1980). Uncertain-frequency detection: Cuing and condition of observation. *Perception and Psychophysics, 28,* 143–149.

Leis-Rossio, B. (1986). *Temporal specificity: Signal detection as a function of temporal position.* PhD thesis, University of Iowa.

Macmillan, N. & Schwartz, M. (1975). A probe-signal investigation of uncertain-frequency detection. *Journal of the Acoustical Society of America, 58,* 1051–1058.

Müller, H. & Findlay, J. (1987). Sensitivity and criterion effects in the spatial cuing of visual attention. *Perception and Psychophysics, 42,* 383–399.

Penner, M. (1972). The effect of payoffs and cue tones on detection of sinusoids of uncertain frequency. *Perception and Psychophysics, 11,* 198–202.

Posner, M. I. & Cohen, Y. (1984). Components of visual orienting. In H. Bouma & D. Bouwhuis (Eds.), *Attention and performance, X,* pp 531–556. Hillsdale, NJ: Erlbaum.

Posner, M. I., Nissen, M. J. & Ogden, W. C. (1978). Attended and unattended processing modes: The role of set for spatial locations. In H. L. Pick & I. J. Saltzman (Eds.), *Modes of perceiving and processing information,* pp 137–157. Hillsdale, NJ: Erlbaum.

Possamaï, C-A. (1986). Composante spatiale de l'attention: Résultats et théories. *Revue Canadienne de Psychologie, 40,* 388–413.

Repp, B. H. (1982). Phonetic trading relations and context effects: New evidence for a phonetic mode of perception. *Psychological Bulletin, 92,* 81–110.

Repp, B. H. (1985). Perceptual coherence of speech: Stability of silence-cued stop consonants. *Journal of Experimental Psychology: Human Perception and Performance, 11,* 799–813.

Rhodes, G. (1987). Auditory attention and the representation of spatial information. *Perception and Psychophysics, 42,* 1–14.

Santon, G. & Scharf, B. (1987). Franco-American differences in the discrimination of DA from GA (Abstract). *Journal of the Acoustical Society of America, 82,* S 118.

Scharf, B. (1970). Critical bands. In J. Tobias (Ed.). *Foundations of modern auditory theory,* Vol. 1, pp. 157–202. New York: Academic.

Scharf, B. (1988). The role of listening in the measurement of hearing. In S. Stephens (Ed.), *Advances in audiology.* Vol. 5, pp. 13–26. Basel: Karger.

Scharf, B., Canévet, G., Possamaï, C-A., & Bonnel, A-M. (1986). Some effects of attention in hearing. Paper presented at the XVIIIth International Congress of Audiology, 24–28 August 1986, Prague.

Scharf, B., Quigley, S., Aoki, C., Peachey, N., & Reeves, A. (1987). Focused auditory attention and frequency selectivity. *Perception and Psychophysics, 42,* 215–223.

Sorkin, R., Pastore, R., & Gilliom (1968). Signal probability and the listening band. *Perception and Psychophysics, 4,* 10–12.

Sperling G. (1984). A unified theory of attention and signal detection. In R. Parasuraman & D. Davies (Eds.), *Varieties of attention,* pp. 103–181. New York: Academic.

Swets, J. & Kristofferson, A. (1970). Attention. *Annual Review of Psychology, 21,* 339–366.

Tanner, W. & Norman, R. (1954). The human use of information: II. Signal detection for the case of an unknown signal parameter. *Transactions of the Institute of Radio Engineers, Professional Group on Information Theory, PGIT, 4,* 222–227.

Ullman, J. & Scharf, B. (1975). *The effect of loudness on attention* (Report). Boston: Northeastern University, Auditory Perception Laboratory.

Yama, M. & Robinson, D. (1982). Comparison of frequency selectivity for the monaural and binaural hearing systems: Evidence from a probe-frequency procedure. *Journal of the Acoustical Society of America, 71,* 694–700.

Static and Dynamic Exertion: A Psychophysical Similarity and Dissimilarity

Joseph C. Stevens

Abstract

At about the same time and apparently quite independently, Borg (see Borg, 1962; Borg & Dahlström, 1959, 1960) and J.C. Stevens and Mack (1959) began studies of how perceived exertion depends on the physical level of performance. Borg and his associates concentrated on dynamic exercise (e.g., cycling), Stevens and his collaborators on static exercise (e.g., handgrip squeezing, curling). Both kinds of exercise turned out to obey the power law of S.S. Stevens (1957, 1975) with very nearly identical exponents (1.6–1.7). This means that perceived exertion always grows as an accelerating function of physical level, so that the closer one approaches maximal exertion the faster the sense of exertion increases – probably helping to protect against straining, and paradoxically enabling greater amounts of work to be done. Perceived exertion also grows over time, given a constant physical level of performance. Here the psychophysical properties of dynamic and static exercise diverge. For example, perceived level of cycling builds up considerably more slowly than does perceived level of handgrip squeezing and, unlike static handgripping, its growth rate depends on the level sustained. These differences show themselves in the slopes of the hyperbolic functions (the so-called Grosse-Lordemann and Müller curves) relating physical level to the duration beyond which the exertion cannot be further sustained.

Introduction

It is obvious from everyday experience that the degree of exertion a person perceives depends both on physical level of exertion and on how long the exertion is continued. Can the degree of perceived exertion be quantitatively related to the physical variables of level and duration of exertion? This question took on considerable interest in about 1960, very shortly afer S.S. Stevens (1957, 1975) stated his claim that perceived magnitudes of several kinds of sensations (e.g., loudness, brightness, saltiness) are related to the physical magnitude (e.g., sound pressure, luminance, molarity) by a power function, i.e.,

$$\psi = k\phi^{\beta}, \quad (1)$$

where ψ is numerically rated by groups of subjects (e.g., using the method of magnitude estimation), ϕ is the level of the physical stimulus, k is a constant of proportionality, and β is the exponent–whose value varies considerably from one sensory modality to another and also within a modality as a function of various stimulus parameters, state of adaptation of the sense organ, and so forth.

Equation 1 can also be written in logarithmic form:

$$\log \psi = \beta \log \phi + \log k. \tag{2}$$

Thus, in log-log coordinates a power function forms a straight line whose slope equals β and whose intercept (up-down location in the coordinates) is indicated by log k.

Perceived Exertion as a Function of Physical Level

Quite independently Borg (see Borg 1962; Borg & Dahlström 1959, 1960) on th eastern side of the Atlantic and J.C. Stevens (see J.C. Stevens & Mack, 1959) on the western side began investigations (much extended as the years went by) on the question of whether perceived exertion also exhibits a power relation. There was one big difference: Borg and his colleagues took interest mainly in dynamic exertion, especially cycling, whereas J.C. Stevens and his colleagues took interest mainly in static exertion, especially handgrip squeezing at first, but later extended here and elsewhere (notably Eisler, 1962) to several other kinds of static work such as foot pressing, forearm curls, finger squeezes, biting. Table 1 lists 12 different studies of various kinds of static and dynamic exertion and the exponents reported.

Unlikely as it might have seemed in advance, the exponents for dynamic and static exertion turned out to be nearly identical (1.6–1.7). The exponent turned out to be virtually constant across the various static tasks. The implication of this number is that perceived exertion is a positively accelerating function of physical level. For example, the average subject estimates that perceived exertion doubles when the physical level increases by only half as much again. The closer and closer one approaches the maximal exertion the more and more rapidly the sensation increases. This relation may serve to protect us from undue strain and to help prevent us from working at very high levels in order to permit a greater work output over time. (See below for a quantitative argument for this assertion.)

The relative invariance of the exponent across so many tasks, static and dynamic (see below for other examples of dynamic tasks), and so many investigators would seem to be one of the basic generalizations in human ergonomics. The notable exceptions have been some studies of apparent heaviness of lifted weights. The wide spread of reported exponents suggests that secondary properties of the weights scaled (e.g., volume, density, even surface color and texture) may have come into play (for an extensive review see Jones, 1986).

Table 1. Power functions for various types of static and dynamic exertion

Researchers	Activity	Methods	Estimated Exponent
Bernyer (1957)	Presses on a hand lever	Halving Doubling	2.0
J. C. Stevens & Mack (1959)	Handgrip	Halving Doubling Magnitude estimation Magnitude production Cross-modality matching	1.7
Girden (see J. C. Stevens & Mack, 1959)	Rotational pressure on a knob	Halving	1.7
Eisler (1962)	Foot presses	Magnitude estimation Magnitude production Cross-modality matching	1.65
Borg (1962)	Cycling	Halving Doubling Ratio estimation Magnitude estimation	1.6
Eisler (1965)	Foot presses	Magnitude estimation	1.5
Bakers & Tenny (1970)	Respiration pressure on a manometer	Magnitude production Magnitude estimation	1.6
van Benthuysen & Stevens (see J. C. Stevens, 1974)	Forearm curls	Magnitude production	1.8
J. C. Stevens (1974)	Finger squeezes	Magnitude production Magnitude estimation	1.6
J. C. Stevens (1974)	Biting	Magnitude production Magnitude estimation	1.6
Cafaralli, Cain, & Stevens (1977)	Cycling and cranking	Magnitude production	1.4
Harver & Kotses (1987)	Static respiratory force	Magnitude production	1.7

After the early work established that the power function describes well the relation between perceived and physical exertion, Borg turned much of his attention to two main topics: (1) individual differences in the absolute perceived magnitude across subjects of different strength, and (2) the physiological correlates of perceived exertion.

In our laboratory the main focus came to be on the question of how perceived exertion changes over time and how physical level and duration trade one for the oher to bring about a constant perceived level. The purpose of the remainder of this paper is to report some of the highlights of our research on this question. As we shall see, it is in this domain that salient psychophysical differences between static and dynamic exertion reveal themselves.

Perceived Exertion (Static) as a Function of Duration

To study the relation between perceived static exertion and its duration (J.C. Stevens & Cain, 1970) we called upon 18 male subjects to squeeze a handgrip dynamometer at various force levels between 22 and 220 N, each held constant over various durations between 4 and 60 s. The task was to estimate (magnitude estimation) perceived exertion experienced at the end of each contraction. There were long rests between contractions to allow for recovery from fatigue. The averages (geometric means) across subjects are shown in two ways: in Fig. 1 as a function of force (duration being the parameter) and in Fig. 2 as a function of duration (force being the parameter). Both ways the data approximate power functions, but the exponents for force (about 1.4) are approximately 2.5 times higher than those for duration (about 0.57). Figure 3 shows the 12 exponents (slopes) as determined by the method of least squares; this shows that there was little or no systematic change of the exponents either as a function of force or of duration. The exponent for force is a little lower than 1.7, probably because of what has been called regression bias (S.S. Stevens & Greenbaum, 1966) associated with relying solely on the method of magnitude estimation in a difficult task. What this means is that people tend to foreshorten whatever dependent variable is in their command, in the present case the number continuum. The effect is to flatten the obtained psychophysical function. The tendency to do this seems to be greater the more difficult the task. The effect of regression on magnitude estimation can, if circumstances permit, be approximately offset by having the subjects adjust the force level in response to numbers given by the experimenter. Now the force level becomes the dependent variable, and the effect is to steepen the obtained psychophysical function. An example of how this works for handgrip was given by J.C. Stevens and Mack (1959).

The important point here, however, is the *relative* sizes of the two exponents. A single equation can now be written to describe how perceived exertion

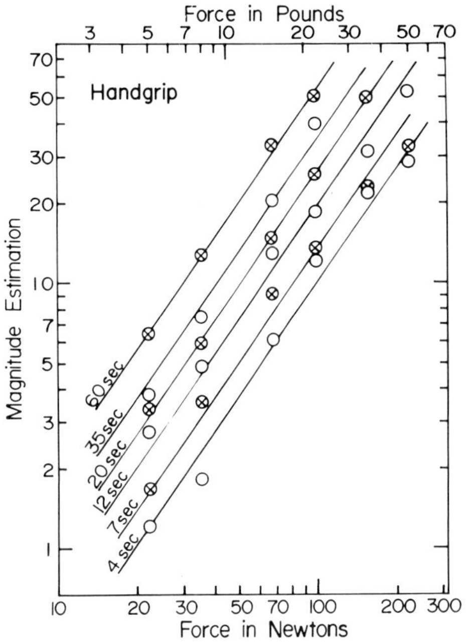

Fig. 1. Perceived terminal exertion (handgrip squeezes) as a function of force as determined by the method of magnitude estimation. The *points* are geometric means, and the *lines* were fitted as a family by the method of least squares under the constraint of a common slope of 1.4 as suggested by Fig. 3. The parameter is the duration of the contraction

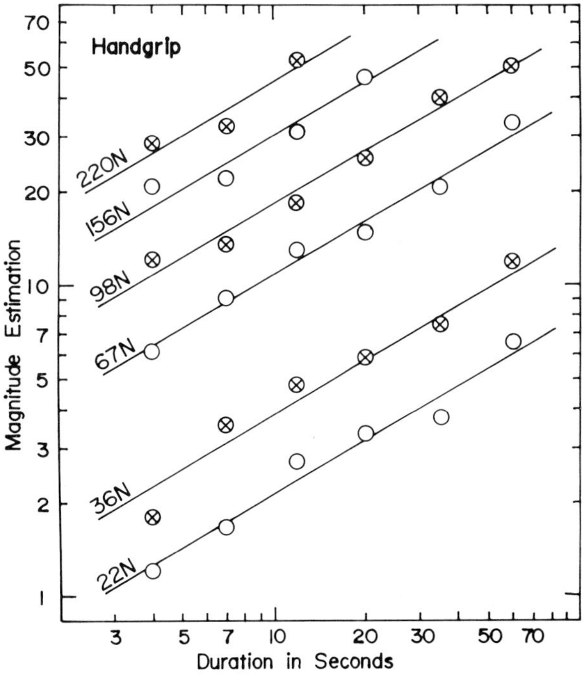

Fig. 2. Perceived terminal exertion (handgrip squeezes) as a function of duration of contraction. The *points* are geometric means, and the *lines* were fitted by the method of least squares under the constraint of a common slope of 0.57 as suggested by Fig. 3. The parameter is the force of the contraction

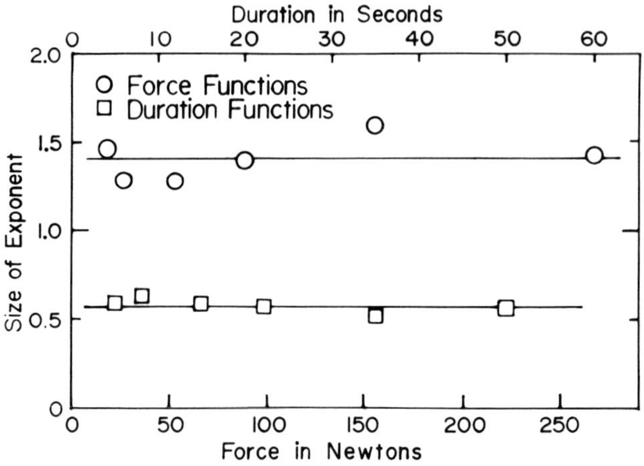

Fig. 3. Exponents as obtained by the method of least squares, operating on geometric means. The *lines* show the average slopes; these averages were used in the construction of Figs. 1, 2

depends on force level and duration t:

$$\psi = k_1 \phi^{1.4} t^{0.57}. \tag{3}$$

If the exponents are increased by about 20% to compensate for the estimated effect of regression this becomes

$$\psi = k_2 \phi^{1.7} t^{0.7}. \tag{4}$$

Both Eqs. 3 and 4 can be used to specify in exactly the same way the combinations of force and duration that will produce a constant degree of perceived terminal exertion (ψ_c). The equation is

$$\psi_c = k_3 \phi^{1.4} t^{0.57} \tag{5}$$

or by rearrangement of terms

$$\psi_c = k_3 \phi^{2.43} t, \tag{6}$$

or

$$t = k_4 / \phi^{2.43}. \tag{7}$$

The important point about Eq. 7 is that it has exactly the same form and very nearly the same exponent as the equation concerning the maximal duration for wich a given force can be sustained, i.e., the so-called endurance function, sometimes called the Grosse-Lordemann and Müller (1937) equation to honor those who apparently first advocated its use. Several investigators have sought to study this relation. Its equation is:

$$t = k_5 / \phi^c. \tag{8}$$

Table 2. Representative time level indices (*c*) for various static and dynamic tasks

Task (static)	*c*	Researchers
Arm, finger, thigh [a]	2.4	Monod & Scherrer (1957)
Arm, leg, trunk [a]	2.1	Rohmert (1960)
Arm [a]	2.5	de Vries (1968)
Handgrip [a]	2.3	Humphreys & Lind (1963)
Handgrip [b]	2.4	J. C. Stevens & Cain (1970)
(Dynamic)		
Cycling (two-limb) [b]	4.4	Cafarelli, Cain, & Stevens (1977)
Cycling (one-limb) [b]	2.9	Cafarelli, Cain, & Stevens (1977)
Cranking (two-limb) [b]	3.6	Cafarelli, Cain, & Stevens (1977)
Cranking (one-limb) [b]	3.0	Cafarelli, Cain, & Stevens (1977)
Cycling (two-limb) [a]	4.7	Tornvall (1963)
Cycling (two-limb) [a]	5.8	Ahlborg (1967)
Repeated handgrip [b]	5.6	J. C. Stevens & Krimsley (1977)

[a] Equal-effort functions, [b] exhaustion functions

A list of exponents *c* is given in Table 2, together with a list of exponents for dynamic exertion, discussed below. The important conclusion here is that duration and force trade one for the other in approximately the same way for both endurance and for any other criterion of constant terminal perceived exertion.

Equation 6 also teaches a lesson, pretty much known instinctively, that to avoid excessive fatigue and strain it is usually better to work at a low level over relatively long durations than at a high level over short durations. Figure 4 is an attempt to illustrate this point quantitatively. The upper contour of the figure shows how much impulse can be expended (i.e., ϕt, a quantity analogous to work in dynamic exertion) at various degrees of constant force applied over time to the point of exhaustion, i.e., the contraction can no longer be maintained at criterion force level. Clearly it is possible to expend much more impulse at low levels of force. This is a common concern of everyday life. The bag of groceries carried from the supermarket may feel quite manageable on leaving, but as time goes by it produces more and more sense of strain and eventually it may come to be intolerable. Fearful that it may become impossible to hold on, a person decides to put the load down to rest. A lighter load it might be possible to carry for all practical purposes indefinitely. Also in Fig. 4 (lower contours) is shown the impulse that can be expended for various degrees of constant terminal perceived exertion. They have, of course, the same general shape as the exhaustion curve. Their generic equation is

$$\phi t = k_6/\phi^{1.43}. \tag{9}$$

In everyday life, of course, things are somewhat more complicated. One may be bored or pressed for time and therefore willing to pay the price of a greater

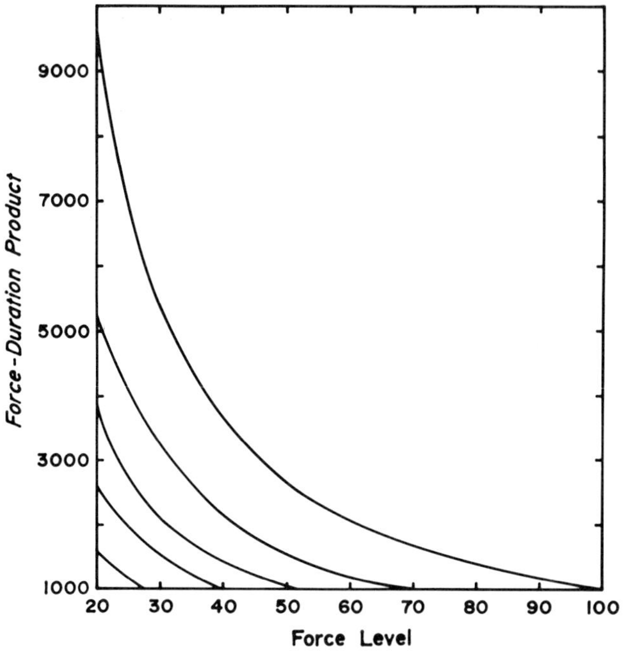

Fig. 4. The functions for handgrip squeezing that relate the impulse expended (product of force and duration) to the point of exhaustion or to the time at which the contraction can no longer be sustained at the criterion force level (*uppermost* function) and the impulse expended up to various constant levels of terminal perceived exertion short of exhaustion

feeling of strain. The serious runner knows how essential it is to pace the speed according to the length of the race. Nevertheless, Fig. 4 provides one kind of general guideline on how to avoid undue strain and exhaustion. And perhaps the very nature of the high exponent (1.7) for the power function that relates apparent exertion to physical exertion serves the purpose of automatically leading a person to work at temperate levels unless there is good reason to do otherwise.

One final word about the measurement of the growth of perceived exertion as a function of duration of static exertion. Jones and Hunter (1983) have presented data that they believe to be at odds with the study described above. They had subjects sustain a contraction of a given level in one arm, up to exhaustion, while attempting at intervals to match the momentary sensation in that arm with a phasic contraction of the other arm. The data have no bearing on the question of how perceived degree of exertion depends on physical level, but they suggest that perceived exertion grows over time at faster rates for stronger loads than for weaker ones, whereas the results presented above suggest a constant growth rate across loads. Unfortunately Jones and Hunter plotted their results in linear rather than log-log coordinates so it is somewhat uncertain whether the two studies do really disagree on this matter. Assuming that the two methods did give different results regarding growth rate, it is unclear why. It could be, for example, that sustained activity in one arm affects the other, e.g., by radiation of muscle activity (when contractions of even relatively small muscle groups are sustained for long periods of time it is impossible to suppress activity in other muscles) or by chemical communication. Either could explain the apparent discrepancy. Jones and Hunter speculate that the method of magnitude estimation is at fault, but if so it seems a remarkable coincidence that our results throughout obey the Grosse-Lordemann and Müller law with the correct exponent.

Perceived Exertion (Dynamic) as a Function of Duration

The measurement of perceived level of dynamic exertion over time is no easy task. This is because the growth rate is much slower than for static tasks, placing considerable strain on a person's memory. Nevertheless, it was possible to glean much information regarding this and other questions using an elaborated version of the paradigm used in the experiment by J. C. Stevens and Cain (1970) in which force and duration of static exertions were both varied in the same test session.

Cafarelli, Cain, and Stevens (1977) called upon ten young male subjects to perform a variety of tasks (cycling with both legs, cycling with only one leg, cranking with both arms, cranking with only one arm) at various power levels and over various durations. These were interspersed among each other in the same test session, and the subject was instructed to make magnitude estimations (on a common scale across tasks) experienced at the end of each work bout. Actually, the subject made two estimates each time (suggested by the work of Ekblom & Goldbarg, 1971), namely the exertion as experienced as coming from the working limbs themselves (local signals) and as experienced as coming from the body core (central signals). It turned out that the local component was always about three times greater than the central component. Given the number of stimulus conditions and the rather long rests required between bouts, it was necessary for each subject to serve in four test sessions; the results for these were tied together by means of a standard bout of two-leg cycling at 100 W for 30 s. This was performed at the start of the session and was assigned the number 10 for its terminal perceived exertion.

The first question concerns how perceived exertion depended on load for each task and every duration (some 20 psychophysical functions). The answer is fairly simple: all conditions approximated power functions, and the exponent averaged 1.43 for cycling and 1.44 for cranking, with little, if any, significant variation across duration or muscle mass (i.e., one or two limbs). This result is remarkably similar to that taken from static exertion. Neither task, nor muscle mass, nor duration seemed to alter the exponent significantly. The exponent is somewhat smaller than estimated in earlier studies (Borg, 1962,

1972). Again, this is probably to be expected as a result of greater regression owing to the difficulty of the task. Be that as it may, it is clear that the perceived levels of both static and dynamic exertion are positively accelerated functions of physical load.

Like the earlier functions of Borg the present ones for cycling (but not cranking) showed a slight upward concavity in log-log coordinates. This concavity disappeared when a small constant was added to all the data, altering the equation of the power function to

$$\psi = k_7 \phi^{\beta} + C. \tag{10}$$

The question arises of whether C is merely a curve-fitting gimmick or whether it has an empirical meaning. It was hypothesized that C might represent the exertion experienced while cycling against zero load. In a secondary experiment we had ten subjects estimate the perceived exertion aroused by several loads, including zero load, over 15-s bouts. When the average estimate of zero load was used as the constant C the data fitted Eq. 10 with excellent precision, thus supporting the hypothesis.

The commonality of the exponent to so many conditions hardly means that these conditions have no bearing on the degree of perceived exertion. Rather, the conditions reveal themselves in the intercepts, or relative up-down positions, of the functions in log-log coordinates. In general, it may be said that, for a given load, perceived exertion was greater for cranking than for cycling, greater for one-limb cycling and cranking than for two-limb, and for a longer duration than for a shorter one. This latter finding, i.e., how perceived exertion grows over time constitutes the next concern of this paper.

In general it may be said that a power function served well to describe the growth of perceived exertion over time. Although the exponent varied significantly across conditions, the most important thing to note is that it is typically very small – on average 0.20 for cycling, 0.34 for cranking. This is on the order of half the corresponding exponent obtained for static handgrip contraction (0.57). In other words, the growth of dynamic sensation is much slower than that of static sensation. For example, to double the estimated perceived exertion of a static handgrip contraction would require a 3.4-fold increase in duration (a little less if regression is taken into account), but to double the estimated perceived exertion of cycling might require (typically) a 32-fold increase in duration (based on an exponent of 0.20). These examples illustrate that although both kinds of sensations grow as negatively accelerating functions of duration, the growth rate for dynamic sensation is in general very much slower than that for static sensation. Thus, though static and dynamic exertion show a remarkable resemblance when it comes to the dependence of perceived exertion on physical exertion, they are quite different in their temporal characteristics. The reasons for the difference may be quite complex, but it seems reasonable to suspect that blood flow has something to do with it since static contractions partially occlude the flow of blood to the contracting muscles.

Table 3. Exponents for perceived exertion as a function of dynamic work duration for cycling and cranking with one and two limbs

Mode[a]	Power (W)							Mean
	25	35	50	75	100	150	200	
Local								
Cycle/2	0.07	–	0.11	0.10	0.17	–	0.26	0.14
Cycle/1	0.21	–	–	0.22	–	0.43	–	0.29
Crank/2	0.29	0.28	0.37	0.42	0.30	–	–	0.33
Crank/1	0.46	–	0.36	0.25	–	–	–	0.36
Central								
Cycle/2	0.16	–	0.29	0.35	0.49	–	0.56	0.37
Cycle/1	0.34	–	–	0.44	–	0.63	–	0.47
Crank/2	0.28	0.27	0.53	0.54	0.48	–	–	0.42
Crank/1	0.35	–	0.43	0.48	–	–	–	0.42
Local plus central								
Cycle/2	0.10	–	0.16	0.17	0.26	–	0.34	0.21
Cycle/1	0.24	–	–	0.27	–	0.47	–	0.33
Crank/2	0.29	0.28	0.41	0.45	0.34	–	–	0.35
Crank/1	0.43	–	0.37	0.31	–	–	–	0.37

[a] 1, 2 refer to the number of limbs used

Table 3 lists the exponents for the 48 psychophysical functions for duration based on the data obtained in this experiment. The main features to be noted (aside from the fact that no one of the 48 exponents is as large as that obtained for static exertion and most are much smaller) are: (a) The duration exponents for cranking were higher than those for cycling, as stated above, even though their exponents with regards load level were practically identical. It would be interesting to know whether various static tasks might also differ in their temporal properties despite their common exponent for force level. So far it appears that only handgrip squeezes have been examined over time. (b) The exponent for cycling increased progressively with load. (c) The exponent for one-limb cycling was considerably higher than that for two-limb cycling; the corresponding picture for cranking is unclear from Table 3.

Thise experiment by Cafarelli et al. is hardly the first to recognize a difference in the temporal porperties of static and dynamic exertion. For example, it has been noted that the maximal level below which a person can apparently exert "indefinitely" is smaller for static than for dynamic exertion (Åstrand & Rodahl, 1970; Monod & Scherrer, 1957; Pandolf & Cain, 1974; Simonson, 1971).

A further difference is the way in which physical load level and duration of exertion trade for each other to produce (a) exhaustion or (b) some constant level of perceived exertion, the same questions raised above concerning static exertion. For both (see Table 2) the results generally conform well to the

Grosse-Lordemann and Müller rule. Although there is considerable variation in the exponent c of the trading function relating load (in watts) to duration (Eq. 8), c is nevertheless in every instance greater for dynamic tasks than for static tasks, reflecting, of course the faster growth of perceived exertion of static exertion over time. It should also be pointed out that since the slopes of the duration functions for cycling and the terminal perceived level of repeated handgrip contractions (which behave like dynamic functions) are level dependent, the exact value of c depends on the particular constant perceived level chosen to construct the trading function. Hence the values of c in Table 2 (dynamic tasks) are merely representative.

The trading functions for dynamic exertion emphasize even more strongly than those for static exertion the wisdom of working at moderate pace unless considerations come into play to dictate otherwise. The amount of work that can be done to exhaustion or to some constant terminal level of perceived effort (analogous to the functions in Fig. 4 and Eq. 8 for impulse of static contractions) have the equation:

$$t = k_8/\phi^{c-1}. \tag{11}$$

Typically the exponent $c-1$ is quite large (e.g., on the order of 3.5 for two-limb cycling), indicating that at a low enough level one can pedal, for all practical purposes, almost indefinitely.

An interesting example of the principle behind Eq. 11 was furnished by Edwards, Melcher, Hesser, Wigertz, and Ekelund (1972). They found that intermittent exertion and resting (cycling) was more effortful than continuous exertion when the two tasks were matched for time-averaged power expenditure.

Finally, it may be pointed out that most everyday tasks involve a static and a dynamic component. When the goal is to minimize perceived exertion it may be important to concentrate on the static component which, although it may seem benign at first, may, because of its rapid growth rate, come to cause great discomfort and limit the amount of work that can be done.

References

Ahlborg, B. (1967). Capacity for prolonged physical exercise in relation to some anthropometric and other data. *Försvarsmedicin, 3,* 194–202.

Åstrand, P. O. & Rodahl, K. (1970). *Textbook of Work Physiology*. New York: McGraw-Hill.

Bernyer, G. (1957). Discussion in symposium: Le problème de la mesure. Proceedings of the 15th International Congress of Psychology (Brussels), pp. 99–100.

Bakers, J. H. M. C. & Tenney, S. M. (1970). The perception of some sensations associated with breathing. *Respiration Physiology, 10,* 85–92.

Borg, G. A. V. (1962). Physical performance and perceived exertion. *Studia Psychologica et Paedagogica, 11,* 1–64.

Borg, G. A. V. (1972). *A ratio scaling method for interindividual comparisons* (Report No. 33). Stockholm: University of Stockholm Institute of Applied Psychology.

Borg, G. A. V. & Dahlström, H. (1959). Psykofysisk undersökning av arbete på cyckelergometer [Psychophysical study of bicycle ergonometer exercise]. *Sartryck ur Nordisk Medicin, 62,* 1383–1392.

Borg, G. A. V. & Dahlström, H. (1960). The perception of muscular work. Publication of the Umeå Research Library, *5,* 1–27.

Cafarelli, E., Cain, W. S., & Stevens, J. C. (1977). Effort of dynamic exercise: Influence of load, duration, and task. *Ergonomics, 20,* 147–158.

deVries, H. A. (1968). Method for evaluation of muscle fatigue and endurance from electromyographic fatigue curves. *American Journal of Physical Medicine, 47,* 125–135.

Edwards, R. H. T., Melcher, A., Hesser, C. M., Wigertz, O., & Ekelund, L.-G. (1972). Physiological correlates of perceived exertion in continuous and intermittent exercise with the same average power output. *European Journal of Clinical Investigation, 2,* 108–114.

Eisler, H. (1962). Subjective scale of force for a large muscle group. *Journal of Experimental Psychology, 64,* 253–257.

Eisler, H. (1965). The ceiling of psychophysical power functions. *American Journal of Psychology, 78,* 506–509.

Ekblom, B. & Goldbarg, A. N. (1971). The influence of physical training and other factors on the subjective rating of perceived exertion. *Acta Physiologica Scandinavica, 83,* 399–406.

Grosse-Lordemann, H. & Müller, E. A. (1937). Der Einfluß der Leistung und der Arbeitsgeschwindigkeit auf das Arbeitsmaximum und den Wirkungsgrad. *Arbeitsphysiologie, 9,* 454–475.

Harver, A. & Kotses, H. (1987). Perception of static respiratory forces in young and old subjects. *Perception & Psychophysics, 41,* 449–454.

Humphreys, P. W. & Lind, A. R. (1963). The blood flow through active and inactive muscles of the forearm during sustained hand-grip contractions. *Journal of Physiology, 166,* 120–135.

Jones, L. A. (1986). Perception of force and weight: Theory and research. *Psychological Bulletin, 10,* 29–42.

Jones, L. A. & Hunter, I. W. (1983). Perceived force in fatiguing isometric contractions. *Perception & Psychophysics, 33,* 369–374.

Monod, H. & Scherrer, J. (1957). Capacité de travail statique d'un groupe musculaire synergique chez l'homme. *Comptes Rendus de la Société de Biologie, 151,* 1358–1362.

Pandolf, K. B. & Cain, W. S. (1974). Constant effort during static and dynamic muscular exercise. *Journal of Motor Behavior, 6,* 101–110.

Rohmert, W. (1960). Ermittlung von Erholungspausen für statische Arbeit des Menschen. *Internationale Zeitschrift für Angewandte Physiologie einschließlich Arbeitsphysiologie, 18,* 123–164.

Simonson, E. (Ed.) (1971). *Physiology of work capacity and fatigue.* Springfield, Illinois: Charles C. Thomas.

Stevens, J. C. (1974). Psychophysical invariances in proprioception. In F. A. Geldard (Ed.), *Cutaneous communication systems and devices* (pp. 73–77). Austin, Texas: Psychonomic Society, pp 73–77.

Stevens, J. C. & Cain, W. S. (1970). Effort in isometric muscular contractions related to force level and duration. *Perception & Psychophysics, 8,* 240–244.

Stevens, J. C. & Krimsley, A. L. (1977). Buildup of fatigue in static work: Role of blood flow. In Borg, G. (Ed.), *Physical work and effort* (pp. 145–155). Oxford: Pergamon Press.

Stevens, J. C. & Mack, J. D. (1959). Scales of apparent force. *Journal of Experimental Psychology, 58,* 405–413.

Stevens, S. S. (1957). On the psychophysical law. *Psychological Review, 64,* 153–181.

Stevens, S. S. (1975). *Psychophysics: Introduction to its perceptual, neural, and social prospects.* New York: Wiley.

Stevens, S. S. & Greenbaum, H. B. (1966). Regression effect in psychophysical judgment. *Perception & Psychophysics, 1,* 439–446.

Tornvall, G. (1963). Assessment of physical capabilities. *Acta Physiologica Scandinavica, 58* (Suppl. 201).

The Study of Individuals in Psychophysical Measurement *

ROBERT TEGHTSOONIAN

Abstract

This paper argues the importance of developing psychophysical scales for the individual observer. Emphasis on group averages has directed attention away from the central issue of estimating the parameters of psychophysical laws for individual subjects. Even the form of the psychophysical function for individuals has been left in doubt. A number of suggestions are made for productive strategies in studying psychophysical scales in individuals and some preliminary findings with these strategies are reported.

Introduction

Throughout the history of research in psychology there has existed a tension between two powerful traditions, one epitomized by Wilhelm Wundt and the other by Sir Francis Galton. Boring (1950) tells us that "Wundt can count as the founder of general psychology – the psychology of the generalized, human, adult, normal mind – whereas Galton is the founder of individual psychology – the psychology of individual differences in human capacity." The former has often been viewed as pure science emphasizing description and generalization, while the latter has been identified with applied science, concerned primarily with finding solutions to practical problems.

There is little doubt that this distinction provides a valid means for classifying two major groups of problems. On the one hand, the study of the basic principles of psychological functioning (such as memory, learning, and perception) as it occurs in all individuals has long been accepted as fundamental in the development of psychology. On the other, the attempt to distinguish one individual from another on the basis of performance on any of a wide variety of performance tests has been at the heart of the mental testing movement since its inception in the work of such pioneers as Galton, Cattell, and Binet. For the most part, students of these separate problem areas have been content

* This paper was prepared with support from a grant from the National Institute of Mental Health, MH 39515.

to pursue their studies with little regard for the ideas of those in the other camp. Testers and clinicians have been preoccupied with assessing the individuality of their clients, whereas students of the basic processes have regarded individual variation as a source of noise in their experiments. That such noise existed at all was traditionally interpreted as meaning only that in a typical experiment there usually remained some relevant variables that were either unknown to or uncontrolled by the experimenter, but that with enough subjects in the various test conditions their effects could be nullified. Relatively little attention has been paid by this group to the possibility that principles of basic importance could emerge from the study of individual differences, and it is the purpose of this paper to suggest briefly several ways in which this general aversion to the study of individuals in psychophysics has been an obstacle to the development of basic descriptive or theoretical principles.

I should say at the outset that the preceding analysis must be regarded as a kind of simplifying caricature and that there have been some important exceptions to the general distaste for the study of individuals exhibited by basic researchers. For example, Borg (1985) has developed and refined over the years a scale for the rating of perceived exertion (RPE), designed (in his words) "as a simple tool to aid a person in estimating and regulating exercise intensity in most kinds of physical activity." That is, the RPE scale was meant for the study of perceived effort in individual subjects. Another approach which seems to focus on the study of individuals is the functional analysis of behavior, as practised by B.F. Skinner and his followers, which is based on the study of single subjects. But this seems to be more the result of a methodological tour de force in identifying and controlling virtually all variables relevant to such outcome behaviors as key-pecking than of a conviction that there are principles that can better be learned from the study of individuals than the study of groups. Perhaps a better example of this latter strategy is to be found in the argument that data averaged over individuals can yield functions which are systematically different in form than the result for any individual subject contributing to such averages. With regard to the form of acquisition functions in learning tasks it has been frequently rediscovered (e.g., Estes, 1956; Teghtsoonian, 1964) that the continuity evident in the averaged data need not be mirrored in the results for any single subject. Indeed, all subjects could show discrete transitions from a no-learning to a learning state and still permit the appearance of continuity in the averaged data. And a similar argument can be made for frequency-of-detection functions and has been used (Stevens, 1972) to argue for a quantum hypothesis of discrimination. Such argument, however, seems to have been more the exception than the rule, and the greater part of contemporary research on the basic processes continues to rely on the use of sizeable samples of subjects for whom average values are calculated. And the increasing reliance on the collection of data by on-line computers and the automated processing of such results for groups of subjects makes it increasingly unlikely that experimenters will ever look at, let alone study, the results obtained for single subjects.

Psychophysics too has followed this general trend. Despite a long tradition of gathering data from single, trained observers, the interest in measuring sensations at levels well above threshold that was stimulated by Stevens' techniques has largely been focussed on averages for groups of subjects. And while Stevens reported that his first studies of magnitude estimation fitted power functions to the data of individual subjects, he was discouraged by the editor of a leading journal from reporting his findings in that form. The change to group averages was made and with few exceptions that technique has persisted. In the rest of this paper I will try to show a variety of ways in which this approach has had unfortunate consequences, and, on the basis of those arguments, urge the adoption of data analysis procedures that focus in individual observers.

Scepticism about the Power Law

It is hard to think of another psychological rule or law for which more supporting evidence has been obtained, yet many continue to doubt the validity of the psychophysical power law. At least one element in this scepticism may be a lack of convincing evidence at the level of the individual observer. In an extensive review of the early literature Marks and Stevens (1966) concluded that for "a large percentage of observers" matching data are well described by power functions. But much of this evidence as well as what has been subsequently published is marred by a number of procedural characteristics which make it easy for the sceptic to remain unconvinced. The number of stimuli employed has been small, frequently in the range from seven to nine and sometimes smaller. Those are not large numbers for determining the precise form of a function. Second, the intervals between stimuli have commonly been large, with steps of 5–10 dB quite common in attempts to scale loudness. Critics have frequently complained that the use of a few widely spaced stimuli makes it easy for the advocate to report data apparently in good agreement with his theoretical preference. Finally, it has been noted that the use of few judgements per stimulus (two is commonly used) leaves the reliability of the points to which a function is fitted in some doubt. (See below for an account of why such doubt is justified even when a given subject's two judgements for a given stimulus are in good agreement.)

To get a more definitive account of the form of the psychophysical function for individual observers, Luce and Mo (1965) undertook a loudness scaling experiment in which they employed 20 intensities closely spaced at 2- to 3-dB steps and obtained 100 judgments per stimulus for each of 12 observers. For the first time it was possible to examine the fine grain of a magnitude estimation function with many closely packed points extending over a large range, with highly reliable point estimates for individual observers. The results provided little comfort for advocates of the power law, or for that matter any other systematic function. Ten of the 12 subjects showed strangely sinuous functions that accelerated in slope, then flattened out, and then became steep again, in patterns that varied mysteriously from one subject to the next. Their finding has now been duplicated frequently (e.g., Eisler & Montgomery, 1974; Green & Luce 1974; Green, Luce & Duncun, 1977; Luce & Green, 1978), and there seems little doubt that this is the characteristic result for this technique.

The puzzle created by the apparent disparity between this result and the findings summarized by Marks and Stevens remains unresolved and, despite its importance, has received very little attention. Proponents of the power law continue to use very widely spaced stimuli, obtain few judgments of each, and assume that departures from linearity (on logarithmic coordinates) are random in nature. But the Luce–Mo findings have not been disconfirmed and, if correct, they imply that the same departures from the power law occur using the traditional procedure but, because of coarse stimulus spacing, go undetected. The important issue is whether the shorter, more convenient, technique conceals from its users information leading to a conclusion quite different from the one that is usually reached. Thus, after more than 30 years of intensive use of the Stevens-inspired scaling procedures, it remains, in my view, uncertain whether the psychophysical power function can be demonstrated convincingly in the judgements of single observers. Later in this paper I will suggest how this puzzle may be resolved.

Reliability of Power Law Exponents

A second set of issues compels us to examine the judgments of individual observers. Even if it is granted that individual functions are well-described by the power law, there remains the question of whether the exponents reported for individuals are reliable. There can be no doubt that such estimates are needed. Increasingly, investigators wish to employ within-subject designs to evaluate the effect of some experimental manipulation. Often, the main question is: What will happen to *this* subject, not just to some group average?

Some may argue that such reliable estimates are already available, and indeed there is some basis for such a belief. Goodness-of-fit measures for the data of individual subjects are typically quite highly (r values in excess of $+0.95$ are quite common in magnitude estimation experiments), suggesting that the "true" function has been estimated with considerable precision. And secondly, test-retest correlations for a group of observers who receive a second session in a magnitude estimation task are frequently high, suggesting that the measured exponent embodies some permanent quality of each observer, namely (it may be argued) the characteristics of the relevant sensory apparatus for that individual. This belief in stable, characteristic exponents measured for individual subjects is widely held, but in my view is more a matter of faith than of evidence.

There are at least two serious flaws in the foregoing arguments. First, the variability in exponents over individuals is embarassingly high. In scaling loudness it is not unusual for a group of 10–15 subjects to yield a range of exponents going from 0.2 to 1.0. It is hard to accept the implication that the auditory systems of these individuals vary so radically. [Indeed, Luce (1972) has argued that the failure of power law exponents to approach invariance over subjects makes it impossible to state units of subjective magnitude in terms that are independent of the observer, and thus that psychophysical measurement must be fundamentally different from physical measurement.] A somewhat different interpretation of the interindividual variability in exponents, that it is a reflection of differences in the use of numbers, not in the characteristics of individual sensory systems, can hardly be more comforting. Reliable estimates of individual eccentricities in number assignment cannot serve any very useful purpose, either theoretical or practical.

A second critical objection to the claim that traditional techniques yield reliable and stable individual exponents is based on the findings reported by Teghtsoonian and Teghtsoonian (1971, 1983) on session-to-session correlations for such exponents. They have shown that session-to-session stability may be readily disrupted by interfering with the observers' ability to recognize or recall the events of the first session while being tested in the second. Either increasing the time between sessions or reassigning the value of the modulus between sessions can have the effect of reducing session-to-session correlations to statistically nonsignificant levels. They have argued that the apparent stability that is often claimed may result from a two-stage process: First-session variability is high because of the many uncontrolled nonsensory factors (such as choice of the first stimulus, modulus self-selection, or initial judgment), and the effect of these factors is maintained through a second session by the operation of memory. In short, they point out that second or third judgments may indeed look like first judgments, but that resemblance may result from the observer's capacity to remember the judgments used earlier and repeat them in subsequent sessions.

If these objections are valid, then it remains the case that even if we assume the power law to hold for individuals, we have yet to develop techniques for the accurate estimation of exponents for individuals. It seems probable that a solution has been slow in coming at least in part because of the excessive reliance that has been placed in the calculation of group averages rather than in a closer study of the judgments made by individuals.

Possible Solutions: The Search for Good Data

It is one thing to urge more attention to the study of individual subjects, but it is not easy to develop research strategies that will make increased attention productive. What follows are some of the approaches to this problem which we are currently exploring in our laboratory and some of the preliminary findings that we have obtained.

One approach to the problem is to obtain for individuals multiple judgments of the same stimuli that are truly independent. The extended sessions employed by Luce and Green and their associates were intended to achieve that goal, and the use of a large number of stimuli may indeed make it more difficult for subjects to resort to mnemonic strategies. Our approach has been to apply the modulus reassignment technique between each pair of ten successive sessions. In doing so, we had two expectations: First, if the variability across individuals typical of traditional magnitude estimation experiments is, as we have proposed, due to a host of procedural variations such as stimulus order, initial stimulus, modulus, and such, then variation within the ten sessions for a given individual where such factors also vary, should be of a comparable magnitude . . . if those ten sessions are truly independent. And that is what we found. Not only do sessions 2, 3, 4, etc., not resemble session 1, but the ten estimates vary widely, in a manner reminiscent of the exponents obtained on a single session for ten different subjects. Second, since the average for each individual of our ten independent estimates would constitute an unbiased estimate of that individual's true value, free of procedural contaminants, the variability over subjects of those mean exponents would be substantially less than is typical of traditional one- or two-session experiments. This expectation too has been supported, although we still need to increase the size of our subject pool. It appears that better estimates of individual exponents exhibit intersubject variability of a much smaller magnitude than has generally been associated with the standard experimental procedure.

One reason why it has been difficult to add to our pool rapidly is that many subjects are either unable to use the number system effectively, or at least require considerable training before their data can be accepted as useable. It seems likely that a large proportion of adult subjects do not or perhaps even cannot use numbers beyond the level of interval or even ordinal scales. Experimenters who do not scrutinize the data of their subjects for evidence that ratio scale judgments are being made risk basing their conclusions on a strange mish-mash of numbers whose meanings vary widely from subject to subject. We have found that it is not unusual to encounter subjects who may on session 1 assign the numbers 10, 15, 25, and 50 to a given set of four intensities, and on the next session with a newly assigned modulus judge those same four intensities as 1000, 1005, 1010, and 1025. In other cases, subjects who have restricted themselves to integral numbers between 1 and 10 in one session reveal those judgments to be simple rank orderings when in a later session working with a larger modulus, those same integral values are just appended to the new modulus, yielding for example 1001, 1002, 1003, etc. Such aberrations can escape undetected when data are collected by on-line computer and only averaged results are studied. If number matching procedures are to be used at all, it seems imperative that subjects be required to undergo some kind of screening procedure (judgments of line length for example) in which modulus shifting is employed before they are used in the main experiment. Even if the subject pool is drawn from populations of college students it seems likely that the uncritical collection of magnitude estimates will yield average values with no legitimate meaning.

Yet another peculiarity afflicting the way in which many subjects use numbers in making psychophysical judgments is the propensity for certain numbers, such as the single digits from 1 to 10, multiples of 10, and certain multiples of 5. This reliance on preferred numbers has been extensively studied by Baird and his associates (e.g., Baird, Lewis & Romer, 1970), and may have more far-reaching effects on the form of magnitude estimation data than has been previously suspected. Although it may lead to occasional distortions in an individual's function when only one or two judgments are made for each intensity, we now have evidence that the hitherto mysterious sinuosity of the Luce-Green repeated-judging procedure is the artifactual outcome of reliance on preferred numbers. Such an outcome leads to a still more radical recommendation than was offered in the preceding paragraph. Although number matching has had great popularity because of its ease of use and the speed with which data can be generated, the problems that it creates may outweigh those advantages. By contrast, cross-modal matches are, as far as we can tell, free of a preferred-value bias, are probably harder to reproduce from memory, and don't seem to be afflicted with the assortment of misunderstandings that subjects commonly impose on instructions to do magnitude estimation. Alternatively, it may be possible to train subjects to avoid the preferred-number bias.

Finally, and perhaps most important, every effort should be made to extend previous attempts to identify and control the sources of variance that can occur in scaling procedures. Range effects, order effects, and first-judgment effects, among others, often add unknown degrees of variability to experimental outcomes. The old rough-and-ready approach that ignored such parameters and relied instead on averaging over subjects simply won't work in the effort to gather accurate and reliable data for individuals.

Acknowledgements. I wish to thank Martha Teghtsoonian for her comments on this paper and for her collaboration in the development of some of the ideas reported here. I also thank Duncan Luce for his helpful comments on an earlier version of this report.

References

Baird, J.C., Lewis, C., & Romer, D. (1970). Relative frequencies of numerical responses in ratio estimation. *Perception & Psychophysics, 8,* 358–362.

Borg, G. (1985). *An introduction to Borg's RPE-scale.* Ithaca, NY: Movement.

Borking, E.G. (1950). *History of experimental psychology* (2nd ed.). New York: Appleton-Century-Crofts.

Eisler, H. & Montgomery, H. (1974). On theoretical and realizable ideal conditions in psychophysics: Magnitude and category scales and their relation. *Perception & Psychophysics, 16,* 157–168.

Estes, W.K. (1956). The problem of inference from curves based on group data. *Psychological Bulletin,* 53, 134–140.

Green, D.M. & Luce R.D. (1974). Variability of magnitude estimates: A timing theory analysis. *Perception & Psychophysics, 15,* 291–300.

Green, D.M., Luce, R.D., & Duncan J.E. (1977). Variability and sequential effects in magnitude production and estimation of auditory intensity. *Perception & Psychophysics, 22,* 450–456.

Luce, R.D. (1972). What sort of measurement is psychophysical measurement? *American Psychologist, 27,* 96–106.

Luce, R.D. & Green, D.M. (1978). Two tests of a neural attention hypothesis for auditory psychophysics. *Perception & Psychophysics, 23,* 363–371.

Luce, R.D. & Mo S.S. (1965) Magnitude estimation of heaviness and loudness by individual subjects. *British Journal of Mathematical and Statistical Psychology, 18,* 159–174.

Marks, L.E. & Stevens, J.C. (1966). Individual brightness functions. *Perception & Psychophysics, 1,* 17–24.

Stevens, S.S. (1972). A neural quantum in sensory discrimination. *Science, 177,* 749–762.

Teghtsoonian, M. & Teghtsoonian, R. (1971). How repeatable are Stevens' power law exponents for individual subjects? *Perception & Psychophysics, 10,* 147–149.

Teghtsoonian, M. & Teghtsoonian, R. (1983). Consistency of individual exponents in cross-modal matching. *Perception & Psychophysics, 33,* 203–214.

Teghtsoonian, R. (1964). One-trial learning directly observed. *Canadian Journal of Psychology, 18,* 304–310.

Physiological and Clinical Aspects

Rating of Perceived Exertion in Children and Adolescents: Clinical Aspects

Oded Bar-Or

Abstract

Rating of perceived exertion (RPE) has been used with healthy children to identify age- and maturation-related differences in perception of exercise intensity. It has also been used in pediatric patients as a means of assessment and as a tool for exercise prescription. Patient groups in whom RPE has been studied include children and adolescents with neuromuscular disease, obesity, anorexia nervosa, and cystic fibrosis. Data suggest that most patients rate an absolute mechanical power as higher than do their healthy peers. This, however, does not denote an "exaggerated" rating in the patients but merely reflects their relatively low maximal aerobic power. Indeed, when exercise intensity is described as a percentage of maximal power output of the exercising muscle group, there is no difference between the RPE of patients and healthy controls.

Attempts have been made to use RPE as a tool for prescription of exercise intensity. Among adults, this seems to be a viable approach. Recent studies from the author's laboratory do not confirm this for children. Obese children could discriminate among exercise intensities prescribed to them as numbers on the RPE scale, but the actual intensities that they selected were different from those expected from the prescription.

Introduction

Most research on rating of perceived exertion (RPE) in healthy people has been conducted with middle-aged and young adults. Relatively little information is available on RPE in children and adolescents. It suggests that people as young as 8–9 years can grasp the RPE concept and validly rate their perceived exercise intensity (as judged from the correlation of RPE to indices of physiological strain). However, there are some age- and maturation-related differences in RPE. The two main differences are that children seem to assign lower ratings at any given physiological strain (Bar-Or, 1977) and that their RPE seems to decrease faster than in adults following aerobic training and acclimatization to heat (Bar-Or & Inbar, 1977). For a recent review, see Bar-Or and Ward (1989).

Similarly, most research on RPE in people with a disease or an illness has focused on adults (e.g., Borg & Linderholm, 1970; Linderholm, 1986; Noble, 1982; Pollock, Jackson & Foster, 1986; Sargeant & Davies, 1977). The purpose of this paper is to review briefly the available information on RPE in pediatric patients and to suggest ideas for future research in this area.

Some information is available on RPE in such diseases as anorexia nervosa (Davies, Fohlin & Thoren, 1980) and cystic fibrosis (Cerny, Cropp & Bye, 1984). The following review will highlight data on obesity and neuromuscular diseases. These are based on studies from the author's laboratory at the Children's Exercise and Nutrition Centre, McMaster University, Hamilton, Canada. The Borg 6–20 category scale (Borg, 1970) was used in all these studies. It was introduced to all subjects in a standard manner, as originally recommended by Gunnar Borg.

Obesity

Obese children are habitually less active than the nonobese. While social isolation and low fitness play major roles in such hypoactivity (e.g., Bar-Or, 1983), one might also argue that obese children perceive a certain degree of physiological strain as inappropriately high. To test this assumption, we recently completed a study (Ward, Blimkie & Bar-Or, 1986) of 43 obese adolescents and 50 healthy controls of both sexes. Subjects performed an "all-out" progressive-continuous protocol on the cycle ergometer. When related to absolute mechanical power, the RPE in the obese girls and boys was about 1.5–2.5 units higher than in the nonobese. However, when exercise intensity was described as percentage of peak mechanical power, the intergroup difference diminished to a nonsignificant 1 RPE unit or less. The above data suggest that the higher rating by mildly to moderately obese adolescents reflects their lower aerobic fitness rather than an inherent "exaggeration" in the way they perceive exercise intensity.

The test-retest reliability of RPE was assessed in two groups of mildly to moderately obese (30%–38% body fat) girls and boys. In one study (Ward & Bar-Or, 1987), a correlation coefficient of 0.92 was found for 11 obese adolescents when data from all submaximal exercise intensities were combined. In the other study (Bar-Or & Ward, 1988), a separate correlation analysis was done at each of four submaximal exercise intensities, performed by twenty 9- to 15-year-old obese subjects. The test-retest correlation coefficient was lowest (0.59) at the lowest exercise level (20% of peak aerobic power). It increased with the rise in intensity, reaching 0.89 at 80% of peak aerobic power. Such reliability coefficients are similar to those found among healthy children and adolescents (Bar-Or & Ward, 1988; Kahle, Ulmer & Rummel, 1977; Komi & Karppi, 1977) and in adults (Carton & Rhodes, 1985). Similarly, the correlation coefficients between RPE and objective indices of exercise intensity such as mechanical power output, percentage of peak power output, and heart rate

Table 1. Validity of RPE in children and adolescents with a disease. Correlation coefficients between RPE and objective indices of effort intensity in girls (*F*) and boys (*M*) with a disease and in a healthy comparison group. Data from Bar-Or & Reed (1986) and from Ward et al. (1986). (Reproduced with permission from Bar-Or & Ward, 1986)

Group	Number of observations	Sex	Limbs	Heart rate	Power	% Peak power
Obesity	21	F	Lower	0.72	0.71	0.84
	12	M	Lower	0.80	0.74	0.84
Muscular dystrophy/atrophy	13	M	Upper	0.69	0.80	0.92
	40	M	Lower	0.72	0.53	0.79
	22	F	Lower	0.69	0.69	0.72
Spina bifida	14	F	Upper	0.82	0.48	0.80
	20	M	Upper	0.58	0.73	0.74
Cerebral palsy	25	F	Lower	0.78	0.66	0.86
	21	M	Lower	0.77	0.12	0.68
Healthy	89	F	Upper	0.78	0.82	0.91
	100	M	Upper	0.81	0.87	0.86
	141	F	Lower	0.69	0.74	0.79
	122	M	Lower	0.82	0.77	0.86

are as high in obese youths as in healthy youths (Table 1; Bar-Or & Ward, 1988). The correlation with percentage of peak power seems to be higher (0.84) than with absolute power or heart rate.

One can conclude, therefore, that mildly to moderately obese children and adolescents rate a given physiological strain similarly to healthy youths. There seem to be no adiposity-related differences in the reliability and validity of RPE in children. These findings are similar to those suggested by Skinner, Hustler, Bergsteinova and Buskirk (1973) for young adults who varied in their adiposity level.

Neuromuscular Diseases

Assuming that a normal process of perceiving exercise intensity requires an intact neuromuscular system, one could argue that, a priori, people with a neurological or muscular disease might rate their perceived exercise intensity differently from healthy individuals. For example, children with muscular dystrophy, atrophy, or spasticity *might be* deprived of normal proprioceptive and other peripheral cues that are needed to inform the sensorium of the exercise intensity. A similar deficiency might occur in patients with diseased joints, such as those with rheumatoid arthritis. Furthermore, many such young patients have been hypoactive since birth, which may have deprived them of some motoric stimuli and experiences enjoyed by their healthy peers. The lack of sufficient stimuli during infancy might affect the way these individuals perceive the intensity of their physical activities in later years.

Understanding the RPE patterns of such patients may help interpret their attitudes toward exertion and their ability and willingness to participate in physical activities. In addition, possible "aberrations" in the RPE of these youngsters may shed light on pathways and mechanisms of exercise perception in the healthy population.

Responses to progressive exercise were monitored in twenty-four 10- to 20-year-old girls and boys with spastic cerebral palsy, muscular dystrophy, muscular atrophy, and spina bifida (Bar-Or & Reed, 1986). All performed arm cranking exercise and some also leg cycling. The RPE of these patients was compared with that of healthy teenagers (50 performing arm exercise and 51 leg exercise). Individual results for arm exercise among the boys are summarized in Figs. 1, 2. As expected, most patients rated their exercise intensity higher than the controls at any given *absolute* mechanical power. However, when plotted against *percentage of peak aerobic power*, the RPE of the patients was well within the 95% confidence range of the healthy controls. A similar

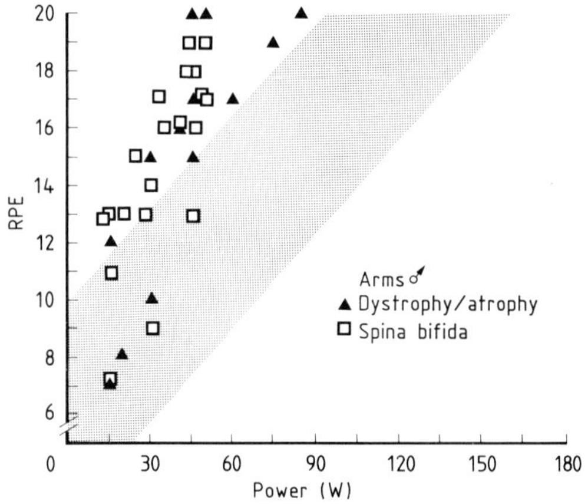

Fig. 1. RPE during arm cranking exercise in relationship to absolute mechanical power in male children and adolescents with neuromuscular disease. *Symbols* represent individual values. The *shaded area* represents 95% confidence range for 51 healthy high school boys. (Reproduced with permission from Bar-Or & Reed, 1986)

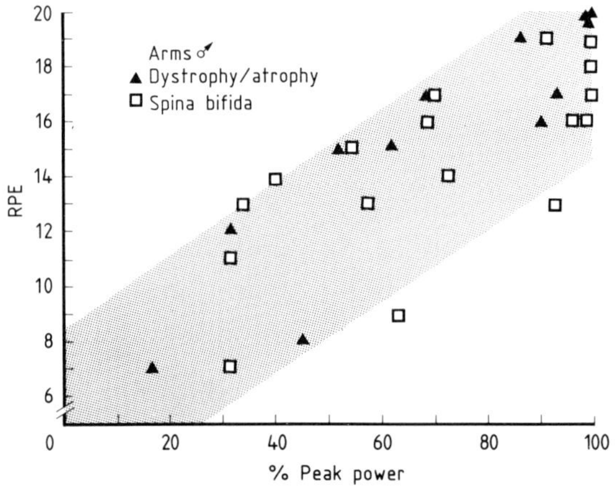

Fig. 2. RPE during arm cranking exercise in relationship to percentage of peak aerobic power in male children and adolescents with neuromuscular disease. Subjects and *symbols* as in Fig. 1. (Reproduced with permission from Bar-Or & Reed, 1986)

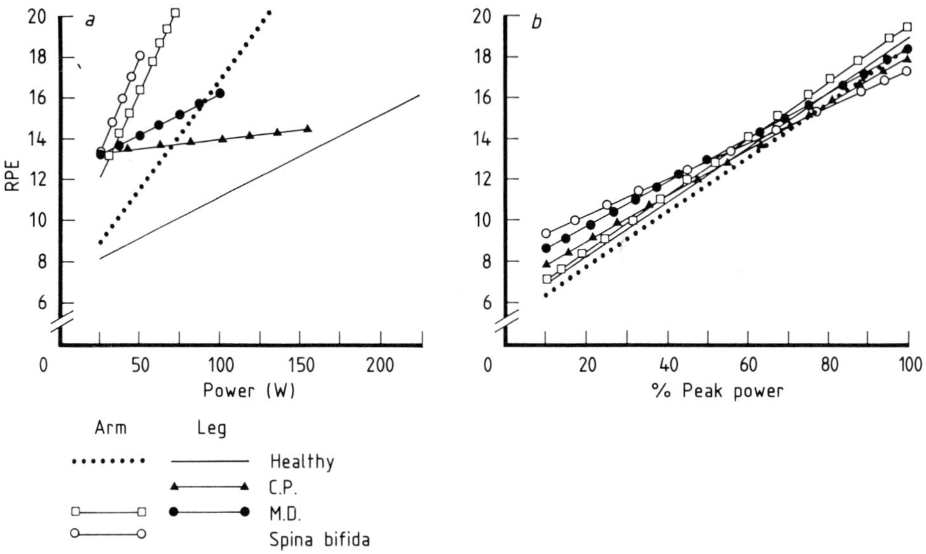

Fig. 3a, b. Regression of RPE on **a** absolute mechanical power and **b** percentage of peak power in teenage males of various diagnostic groups. *C.P.*, Cerebral palsy; *M.D.*, muscular dystrophy. (Reproduced with permission from Bar-Or & Reed, 1986)

pattern was evident for leg exercise and for girls. Figure 3 summarizes the regression line of RPE on mechanical power for each of the diagnostic groups. When RPE is plotted against absolute mechanical power (Fig. 3a), both the intercepts and the slopes are quite different from each other, with the patients generally having a higher RPE than the healthy controls at any given power. In contrast, when RPE is plotted against percentage of peak power (Fig. 3b), the regression lines align closer to each other, irrespective of the diagnosis or of the muscle group that was used for exercise.

The above regularity in RPE, irrespective of the clinical condition of the muscles of these children, is in line with the notion of "muscle matching," as described by Cafarelli (1982) for healthy adults. It is also suggestive of a central feedforward loop that informs the sensory cortex about the intensity of exertion that the muscles are about to perform. Such a loop does not require peripheral feedback cues from the exercising muscles. One cannot, however, exclude the possibility that some peripheral cues are delivered from the respiratory muscles, pleura, or other elements in the respiratory apparatus.

Use of RPE in Exercise Prescription

Ideally, children should be allowed and encouraged to exercise as part of play and recreation, in a spontaneous, nonregimented manner. Many healthy children will perform sufficient exercise when left on their own. However, other

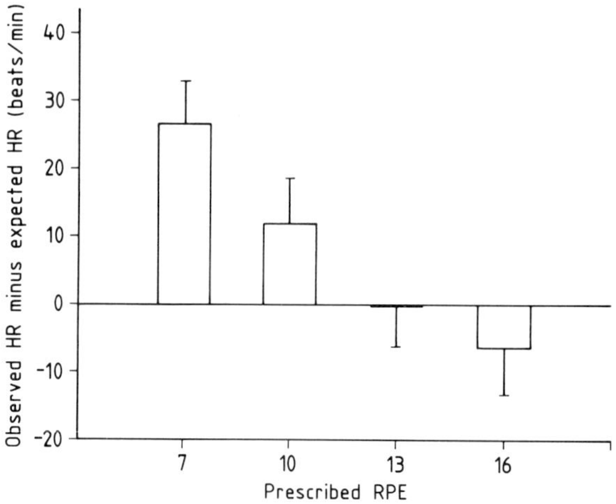

Fig. 4. Accuracy of performing walking/running tasks on the track, as prescribed by RPE. Mean values for the difference in heart rate (*HR*) between that observed during the prescribed task and that obtained in a preliminary cycling session when the RPE scale was taught. Twenty 9- to 15-year-old obese children took part. *Vertical lines* denote 1 SEM. (Reproduced with permission from Bar-Or & Ward, 1989)

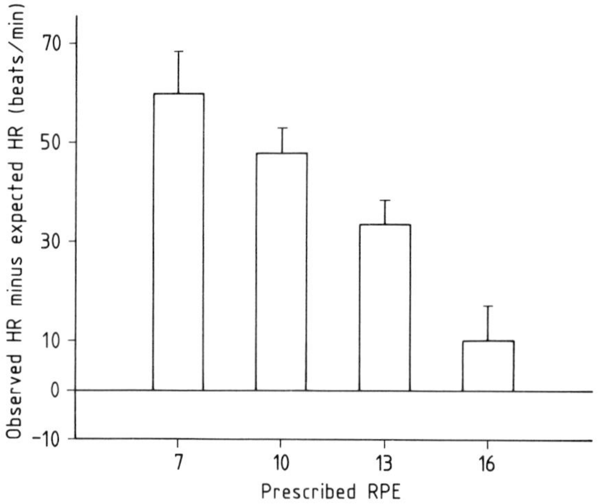

Fig. 5. Accuracy of performing cycling tasks in the laboratory, as prescribed by RPE. Variables, subjects, and notation are as in Fig. 4. (Reproduced with permission from Bar-Or & Ward, 1989)

children, particularly those with a physical or mental disability or those who are socially maladjusted will often need an exercise prescription. Such a prescription must include information on the type, intensity, frequency, and duration of the activities. While the notions of type of exercise, frequency of bouts, and duration of each bout are relatively easy to convey to a child – directly or via the parent – it is hard to explain the notion of specific intensities. The use of heart rate as a gauge of physiological strain, commonly practiced with adults, is less feasible for children. (No published studies are available to us on this topic. The above statement is based on the author's observations at the Wingate Institute, Israel.)

Several authors have evaluated the use of RPE as a means for exercise prescription in adults (e.g., Birk & Birk, 1987; Chow & Wilmore, 1984; Maresh

& Noble 1984; Pollock et al., 1986). Some of these studies suggest that the method is useful, while others are equivocal. The author's laboratory is currently performing a project to evaluate the usefulness of the RPE as a means for prescribing *exercise intensity* to children and adolescents with a disability. In a phase just completed (Ward & Bar-Or, 1987), twenty 9- to 15-year-old girls and boys with mild to moderate obesity were taught the RPE concept through a duplicate multiload cycle ergometer test. At the end of each 2-min load they rated their RPE using the Borg 6–20 category scale and their heart rates were measured. An RPE on heart rate "calibration line" was thus produced for each child. The children were subsequently given four-stage exercise prescriptions, using numbers on the RPE scale to describe each intensity. One prescription was done for a cycle ergometer and the other for an outdoor track (walking/jogging tasks). Results suggested that these children could discriminate among the four prescribed intensities (presented as 7, 10, 13, and 16 on the scale) on the ergometer and, to a lesser extent, on the track. However, when children were instructed to exercise at a particular RPE, they chose activities that were more intense than the prescribed ones as indicated by heart rate during the exercise and the calibration session (Figs. 4, 5). Exaggeration in the chosen effort was particularly evident for all four track tasks. On the ergometer, the subjects exaggerated in their choice of exertion for the RPE = 7 and RPE = 10 tasks. Their choice of exertion for the RPE = 13 task was very similar to the expected intensity, while that for the PRE = 16 was lower than expected. It was thus concluded that these obese children could discriminate among prescribed exercise intensities but their *range of chosen intensities* was well below the expected. It is apparent that some improvement is needed in teaching children the RPE concept. One approach is to let the child repeatedly experience the extremes on the Borg scale. Such "anchoring" of the extremes may eventually allow the children a greater range of intensities to choose from. In a phase just begun we are evaluating the usefulness of the RPE as a means for exercise prescription for wheelchair-bound children and young adults.

Needs for Future Research

The above data represent only part of the spectrum of issues in which RPE is of relevance to pediatric populations. Additional areas that this author would like to see studies include:

1. Differentiated rating of "leg effort," "chest effort," and "general effort." Such information is of particular interest in patients whose effort may be limited by lung function (e.g., cystic fibrosis), leg function (e.g., muscular dystrophy), or both lung and leg function (e.g., advanced stages of anorexia nervosa, severe undernutrition in cystic fibrosis).
2. The possible usefulness of RPE as a diagnostic tool to identify children who *cannot* exercise as opposed to those who can but *will not* exert themselves.

3. Methods of assessing RPE in children younger than 8 years, or who are mentally retarded and cannot learn the concept of a numerical rating scale. (For example, could scales based on pictures or other symbols be better understood by such children?)
4. Refined methods of teaching the RPE concept for the purpose of using it in exercise prescription.
5. Further exploring the use of children with a disease to study the mechanisms involved in perceiving exertion (e.g., patients with McArdle's disease whose lactate production is deficient).

Acknowledgments. Several people have been instrumental in generating and analyzing the data at the Children's Exercise and Nutrition Centre. The author is particularly indebted to Carol Balsilie, B.P.E., Michael Browne, B.P.E., R.N., Leslie McGillis, B.Sc., Dianne Moroz, M.Sc., Sherry Reed, B.P.E., Karen Smith, M.D., and Dianne Ward, Ph.D.

References

Bar-Or, O. (1977). Age-related changes in exercise perception. In G. Borg (ed.), *Physical work and effort* (pp. 255–266). Oxford: Pergamon.

Bar-Or, O. (1983). *Pediatric sports medicine for the practitioner*. New York: Springer.

Bar-Or, O. & Inbar, O. (1977). Relationship between perceptual and physiological changes during heat acclimatization in 8–10 year old boys. In H. Lavallée & R.J. Shepard (Eds.), *Frontiers of activity and child health* (pp. 205–214). Québec: Pélican.

Bar-Or, O. & Reed S.L. (1986). Rating of perceived exertion of adolescents with neuromuscular disease. In G. Borg & D. Ottoson (Eds.), *The perception of exertion in physical work* (pp. 137–148). London: MacMillan.

Bar-Or, O. & Ward, D.S. (1989). Rating of perceived exertion in children. In O. Bar-Or (Ed.), *Advances in pediatric sport sciences,* Vol. 3. Champaign, Il: Human Kinetics.

Birk, T.J. & Birk, C.A. (1987). Use of perceived exertion for exercise prescription. *Sports Medicine, 4,* 1–8.

Borg, G. (1970). Perceived exertion as an indicator of somatic stress. *Scandinavian Journal of Rehabilitation Medicine, 2–3,* 92–98.

Borg, G. & Linderholm, H. (1970). Exercise performance and perceived exertion in patients with coronary insufficiency, arterial hypertension and vasoregulatory asthenia. *Acta Medica Scandinavica, 187,* 17–26.

Cafarelli, E. (1982). Peripheral contribution to the perception of effort. *Medicine and Science in Sports and Exercise, 14,* 382–389.

Carton, R.L. & Rhodes, E.C. (1985). A critical review of the literature on rating scales for perceived exertion. *Sports Medicine, 2,* 198–222.

Cerny, F.J., Cropp, G.J.A., & Bye, M.R. (1984). Hospital therapy improves exercise tolerance and lung function in cystic fibrosis. *American Journal of Diseases of Children, 138,* 261–265.

Chow, R.J. & Wilmore, J.H. (1984). The regulation of exercise intensity by ratings of perceived exertion. *Journal of Cardiac Rehabilitation, 4,* 382–387.

Davies, C.T.M., Fohlin, L. & Thorén, C. (1980). Perception of exertion in anorexia nervosa patients. In K. Berg & B.O. Eriksson (Eds.), *Children and exercise.* IX (pp. 327–332). Baltimore: University Park Press.

Kahle, C.J., Ulmer, H.J. & Rummel, L. (1977). The reproducibility of Borg's RPE scale of female pupils from 7 to 11 years of age (abstract). *Pflugers Archiv. European Journal of Physiology,* (Suppl. 368), 26

Komi, P.V. & Karppi, S.-L. (1977). Genetic and environmental variation in perceived exertion and heart rate during bicycle ergometer work. In G. Borg (Ed.), *Physical work and effort* (pp. 91–99). Oxford: Pergamon.

Linderholm, H. (1986). Perceived exertion during exercise in the discrimination between circulatory and pulmonary disorders. In G. Borg & D. Ottoson (Eds.), *The perception of exertion in physical work* (pp. 177–190). London: Macmillan.

Maresh, C.M. & Noble, B.J. (1984). Utilization of perceived ratings during exercise testing and training. In L.K. Hall (Ed.), *Cardiac rehabilitation: exercise testing and prescription* (pp. 153–173). Amsterdam: Spectrum.

Noble, B.J. (1982). Clinical applications of perceived exertion. *Medicine and Science in Sports, and Exercise, 14,* 406–411.

Pollock, M.L., Jackson, A.S., & Foster, C. (1986). The use of the perception scale for exercise prescription. In G. Borg & D. Ottoson (Eds.), *The perception of exertion in physical work* (pp. 161–176). London: Macmillan.

Sargeant, A.J. & Davies, C.T.M. (1977). Perceived exertion of dynamic exercise in normal subjects and patients following leg injury. In G. Borg (Ed.) *Physical work and effort* (pp. 345–355). Oxford: Pergamon.

Skinner, J. S., Hustler, R., Bergsteinova, V., & Buskirk, E.R. (1973). Perception of effort during different types of exercise and under different environmental conditions. *Medicine and Science in Sports, 5,* 110–115.

Ward, D.S. & Bar-Or, O. (1987). Usefulness of RPE scale for exercise prescription with obese youth (abstract). *Medicine and Science in Sports and Exercise, 19,* 15.

Ward, D.S., Blimkie, C.J.R., & Bar-Or, O. (1986). Rating of perceived exertion in obese adolescents (abstract). *Medicine and Science in Sports and Exercise, 18,* 72.

Comparison of Rated Perceived Exertion and Constant Effort During Cycling Exercise

WILLIAM A. LATZKA, STEPHEN R. MUZA, LESLIE LEVINE, and KENT B. PANDOLF

Abstract

Overall sense of effort during muscular exercise has been described as having local and central components. Both components are thought to be involved in the determination of rated perceived exertion (RPE) and constant effort (CE) during physical exercise; however, RPE and CE power output have not been directly compared. The present study examined the relationship between RPE during steady-state exercise and power output during CE exercise in ten male volunteers. Steady-state exercise (~60% peak oxygen uptake, $\dot{V}O_2$) and CE exercise (initial exercise intensity ~70% peak $\dot{V}O_2$) were each performed for 30 min on a cycle ergometer. During steady-state exercise, RPE increased as the exercise duration increased ($p < .05$). During CE exercise, power output (W) decreased as exercise duration increased ($p < .05$). When W during CE exercise was plotted as a function of RPE during steady-state exercise for identical time intervals, the following linear relationship ($y = bx + a$) was obtained: $W = b \times \text{RPE} + a$ ($r = -.47$) where a is 196.6 and b is -3.14 ($p < .10$). These findings indicate that RPE and CE power output are related and, when studied collectively, may someday help explain the mechanism(s) responsible for the buildup of muscular fatigue during physical exercise.

Introduction

When an individual maintains the force of a static muscular contraction or the power output of a dynamic exercise task such as cycling at a constant level over time, he or she will experience a continuous increase in the perceived effort (Pandolf, Cafarelli, Noble & Metz, 1972; Stevens & Cain, 1970). The magnitude estimation method (Cain & Stevens, 1971; Stevens & Krimsley, 1977) is used extensively to study the perception of effort during constant force or power output exercise such as static handgrip contractions while rated perceived exertion (RPE) (Pandolf, Billings, Drolet, Pimental & Sawka, 1984; Pandolf et al., 1972) is generally employed to evaluate effort sense during dynamic cycling exercise.

In contrast, when an individual maintains a static muscular contraction or dynamic cycling task at constant level of effort, he or she will decrease the force level or power output, respectively, over time (Cain & Stevens, 1971; Pandolf & Cain, 1974). After a very brief exposure to the experimental stimulus (force level or power output), subjects are instructed to adjust the force level or power output so that it never feels any more or less difficult than it had initially. Therefore, sensation of effort is maintained constant.

For static handgrip contractions, two exponential terms were needed to describe the reduction in force level when sense of effort was maintained constant (Cain, 1973, 1977; Cain & Stevens, 1971, 1973; Pandolf & Cain, 1974; Stevens & Krimsley, 1977). Cain and Stevens (1971) speculated that the need for two exponential terms may be related to two distinguishable processes such as the activity of mechanoreceptors in tendons and skin, and the metabolic activity associated with the exercise. For cycling exercise, a two-component exponential function was also necessary to explain the decline in power output with maintenance of constant effort, and they speculated that this function might involve local and central components. The local or peripheral component is thought to involve sensations or feelings from the exercising muscles and/or joints, while the central component is associated primarily with cardiopulmonary sensations (Pandolf, 1977, 1978, 1982; Pandolf & Cain, 1974).

Rated perceived exertion (RPE) and constant effort (CE) have not been directly compared during muscular exercise. The purpose of this study is to compare RPE and CE during 30 min of cycling exercise. Since both RPE and CE are thought to involve local and central components, it was hypothesized that the experimental observations from these two psychophysical techniques would be related.

Methods

Subjects

Ten male volunteers served as test subjects in this study. Each subject was medically cleared prior to participation, and after being informed of the potential risks that were involved with this study, signed a statement of informed consent. Their physical characteristics are presented in Table 1. The subjects were normally distributed with respect to age, height, weight, and peak oxygen uptake ($\dot{V}O_2$). Most of the subjects were above average in aerobic fitness; however, none of the subjects was a trained cyclist. Peak oxygen uptake was measured in a sitting, semisupine position on a cycle ergometer and has been shown to be about 15%–20% less than the maximal oxygen uptake during running to volitional exhaustion on a treadmill (Åstrand & Rodahl, 1986).

Protocol

Initially, the subjects were familiarized with the testing procedures and the mode of exercise to be used throughout this investigation. All tests were

Table 1. Physical characteristics of the subjects

Subject	Age (years)	Height (cm)	Weight (kg)	Peak $\dot{V}O_2$ (ml kg^{-1} min^{-1})
1	22	173.0	73.2	50.7
2	28	185.4	71.6	56.5
3	21	165.5	62.0	43.5
4	24	172.0	69.6	42.6
5	21	171.3	62.5	42.4
6	22	177.5	76.4	48.9
7	34	175.2	82.8	47.8
8	21	173.5	67.2	34.6
9	19	159.2	66.1	53.2
10	23	175.5	71.6	49.7
Mean	23.5	172.8	70.3	47.0
SD	4.4	7.0	6.4	6.3

conducted in a comfortable environment (~22°–25° C). Subjects wore shorts and tennis shoes during these tests.

All testing for each subject was completed within 1 week of the first test day to minimize any training effect. The actual tests were conducted over 2 days.

Anthropometric measurements and peak oxygen uptake were evaluated on the first morning. Each subject's peak oxygen uptake was determined using a progressive intensity, continuous exercise protocol to volitional exhaustion on a modified cycle ergometer. Before the peak aerobic power test, we predicted the subject's maximal oxygen uptake based on his heart rate response to 5 min cycling at 100 W using the nomogram developed by Åstrand (1960). Initially, each subject completed a 3-min warm-up exercise at 90 W. The exercise intensity was then increased to approximately 50% of each individual's estimated maximal exercise intensity and subsequently increased 30 W every 2 min until the subject attained volitional exhaustion. The criteria (Kamon & Pandolf, 1972) employed for the determination of peak oxygen uptake included the leveling off or decrease of the oxygen uptake with increased exercise intensity (an increase in $\dot{V}O_2$ of less than 150 ml/min for a 30-W increment).

That afternoon (about 4 h later) the CE exercise test was conducted. This consisted of a 3-min warm-up at 90 W followed by 30 min exercise. During the first minute, each subject maintained a power output that elicited 70% of his peak oxygen uptake. After 1 min exercise at this intensity, the subject adjusted the load selector dial so that he would maintain the prevailing level of effort constant over the remaining 29 min exercise. Subjects were instructed that, during the initial minute, they were to take special notice of their sense of the effort (i.e., feelings from the exercising muscles and joints, and cardiopulmonary sensations such as heart rate and ventilation) required to maintain the initial power output. Subjects were reminded of the instructions periodically throughout the exercise period. These same procedures for CE cycling exercise have been previously reported (Pandolf & Cain, 1974).

The steady-state exercise test was administered on the second test day. Again, this consisted of a 3-min warm-up at 90 W, followed immediately by 30 min exercise at 60% of each individual's peak oxygen uptake. During this test the subject was asked to rate his exertion as perceived during the exercise. Every 2 min the overall perceived exertion was rated using the classic 15-point Borg scale (Borg, 1970).

All exercise tests were conducted on a Collins electronically braked cycle ergometer with the subjects seated in a semisupine position cycling at 60 rpm. Power output was continuously monitored and recorded. Respiratory parameters – oxygen uptake ($\dot{V}O_2$), carbon dioxide production ($\dot{V}CO_2$), and pulmonary ventilation ($\dot{V}_E$) – were measured by open-circuit spirometry using a Sensor Medics Metabolic cart. The respiratory parameters were averaged over 15 s during peak oxygen uptake tests and over 30 s during the CE and steady-state exercise tests. Heart rate was recorded from chest electrodes (CM-5) every 2 min on an electrocardiograph.

Statistical Treatment

Data from the first 25 min of steady-state and CE exercise were used for statistical analysis in order to avoid any extraneous psychological factors that may have affected the results because of the subjects knowing that the exercise bout was near completion. However, the data points for the full 30 min are shown in Figs. 1–3, although the final data points are shown disconnected from the first 25 min. The means and standard deviations were used to describe these data. Analyses of variance with repeated measures were employed to determine significant differences of the physiological variables between the CE and steady-state exercise tests and of these same variables over time. Tukey's post hoc comparison was used to determine where the differences had occurred. Individual regression equations were developed for each subject using power output during CE as the independent variable and RPE during steady-state exercise as the dependent variable, both measured over the same time intervals in each exercise. Statistical significance was accepted at the $p < .05$ level.

Results

Steady-State Exercise

The means and standard deviations for the overall RPE during steady-state exercise are shown in Fig. 1. As stated previously, during this test the power output was held constant at 60% of the individual's peak oxygen uptake. Overall RPE increased ($p < .05$) over the duration of the steady-state exercise. When RPE was plotted as a function of time, there was a significant positive correlation ($r = .53$). Tukey's post hoc test indicated that RPE after 15 min exercise did not change significantly ($p > .05$), suggesting that RPE started to plateau after 15 min of steady-state exercise.

CE Exercise

The means and standard deviations for power output during CE exercise are shown in Fig. 2. Power output decreased significantly ($p < .05$) during the first 5 min exercise; however, the post hoc test revealed that there were no significant differences ($p > .05$) in power output after 5 min CE exercise.

Physiological Responses

The means and standard deviations for the metabolic and cardiorespiratory responses to CE and steady-state exercise are compared in Fig. 3. The $\dot{V}O_2$ increased ($p < .05$) during the first 5 min of both exercise tests with no significant differences ($p > .05$) observed after this time period during either the CE or steady-state exercise tests. Also, there were no significant differences ($p > .05$) between CE and steady-state exercise after 5 min exercise. These same response patterns and statistical observations were observed for heart rate and pulmonary ventilation.

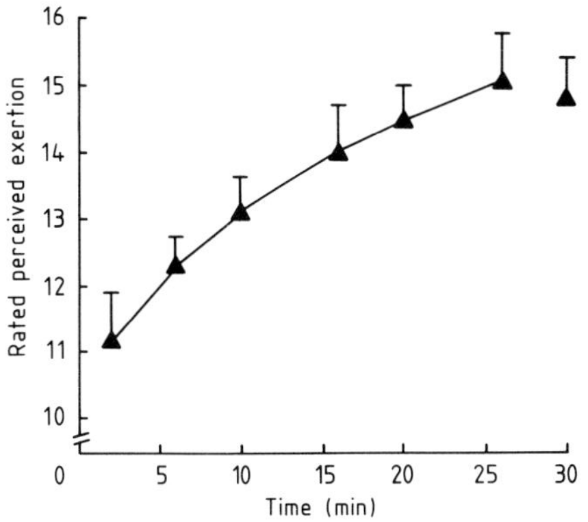

Fig. 1. Rated perceived exertion (RPE) during steady-state exercise of 30 min duration at 60% peak oxygen uptake. Values are mean ± SD

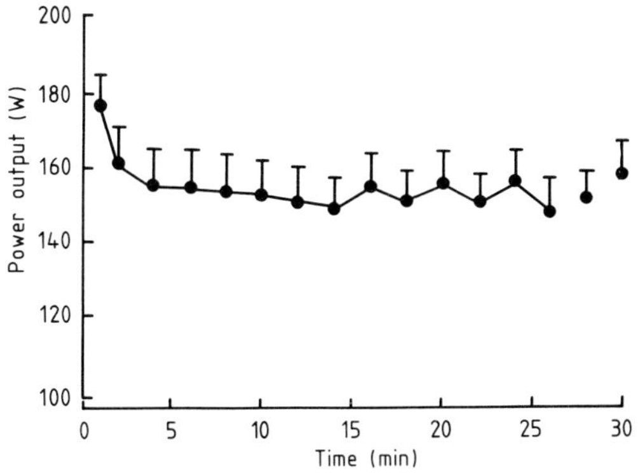

Fig. 2. Power output during CE exercise of 30 min duration. Values are mean ± SD

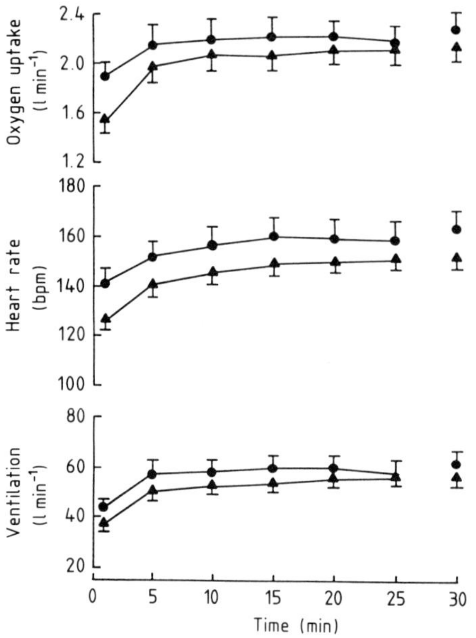

Fig. 3. Comparison of pulmonary ventilation, heart rate, and oxygen uptake during CE (●) and steady-state exercise (▲). Values are mean ± SD

Comparison Between Perceptual Responses During CE and Steady-State Exercise

The individual regression lines shown in Fig. 4 show the relationship between each individual's power output during CE exercise and RPE during steady-state exercise in identical time intervals. It was important to analyze the individual responses because each individual had his own unique response to the scaling or CE and/or RPE. Therefore, intersubject variability was minimized by plotting individual relationships. However, there was still large variability, as can be seen from this figure. Only six of the ten subjects responded as we hypothesized; that is, when the power output during CE exercise was plotted as a function of RPE during steady-state exercise there was an inverse relationship. The linear equation in Fig. 5 represents the mean regression equation for the ten subjects ($r = -.47$, $p < .10$). However, when the four subjects who did not respond as hypothesized were deleted from this analysis, there was a statistically significant relationship ($r = -.54$, $p < .05$).

Discussion

Several reports (Pandolf et al., 1972, 1984) indicate for prolonged physical exercise that a perceptual steady state in terms of effort sense is reached in a far different fashion from a physiological steady state; and, if at all, after a

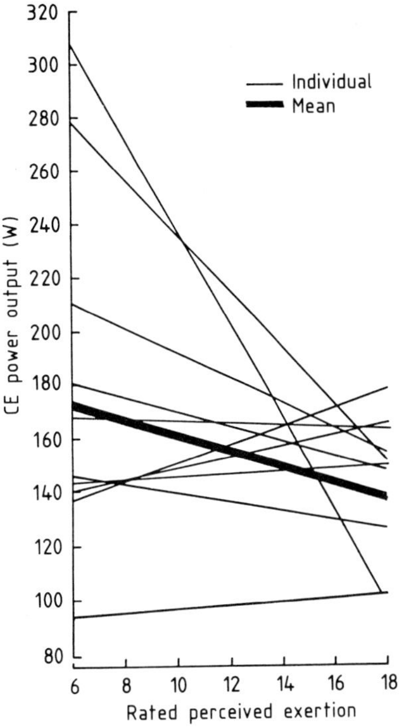

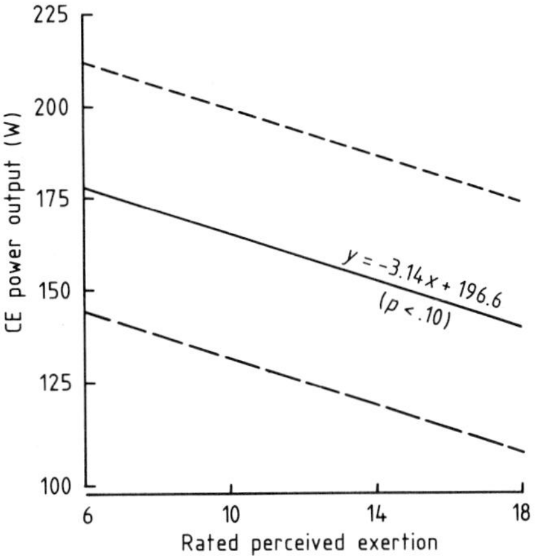

Fig. 5. The mean regression equation for the ten subjects of power output during CE exercise on RPE during steady-state exercise at identical time intervals

Fig. 4. Comparison of the relationship between each individual's power output during CE exercise and RPE during steady-state exercise at identical time intervals

comparably much longer time period. Pandolf et al. (1972) showed that the undifferentiated RPE progressively increased during constant intensity cycling exercise of 30 min duration at ~69% peak oxygen uptake with little indication of a steady-state value being approached at the end of exercise. In a more recent study which contrasted arm cranking to cycling exercise at ~60% peak oxygen uptake, none of the differentiated RPE (local, central, and overall) for upper body exercise reached steady-state levels, even after 60 min (Pandolf et al., 1984). Only the central and overall RPE for lower body exercise appeared to reach near steady-state values, but after 40 min (Pandolf et al., 1984). The present physiological and RPE observations generally support the contention that a perceptual steady state is attained much later during prolonged cycling exercise than a physiological steady state. Thus, some authors (Pandolf et al., 1972) conclude that RPE during prolonged exercise may initially indicate the perceptual adaptation to the exercise task, but later denote the perception of fatigue not exertion.

To our knowledge, only one study has reported observations concerning constant effort during cycling exercise at initial power outputs that ranged from 30% to 85% of peak oxygen uptake (Pandolf & Cain, 1974). At an initial power output equivalent to ~70% peak oxygen uptake, Pandolf and Cain (1974) documented a mean 67 W reduction in power output over 12 min as sense of effort was maintained constant. In the present study at the same initial power output (~70% peak oxygen uptake), a mean reduction in power output of 30 W was observed over a 25-min period during which effort sense was maintained constant. While the overall reduction in power output during CE exercise was less in the present experiments than in those of Pandolf and Cain (1974), the CE curves derived from these two experiments appear to be similar. Further, both experiments support the contention that most of the decline in power output as effort sense is maintained constant occurs during the first few minutes of exercise. Nevertheless, several authors (Cain, 1977; Pandolf & Cain, 1974) suggest the CE procedure as a way to evaluate the subjective side of fatigue and/or the mechanisms responsible for the buildup of fatigue.

Since the decrease in power output with CE and the increase in RPE during prolonged exercise are both thought to be related to the buildup of fatigue, it is also possible that these two responses are related to some degree. In the present study, individual regression lines relating each individual's power output during CE exercise to RPE during steady-state exercise at identical time intervals showed a wide range of responses. However, there was a tendency for individuals with higher initial power outputs during CE exercise to have steeper slopes on the regression of power output on RPE during steady-state exercise, which is in concert with what might be hypothesized from the literature (Cain, 1977; Pandolf and Cain, 1974). While the relationship between power output during CE exercise and RPE during steady-state exercise bordered on being statistically significant ($r = -.47$, $p < .10$) for all ten subjects, this same relationship including only those subjects ($n = 6$) who demonstrated the hypothesized inverse association was significant ($r = -.54$, $p < .05$). The accountable variance (r^2) in common between the power output during CE and RPE during steady-state exercise in these experiments was approximately 30%.

Overall sense of effort during cycling exercise at a constant intensity has been described to have local and central components (Cafarelli, 1977; Pandolf, 1977, 1978, 1982, 1986; Robertson, 1982). The local of peripheral component is thought to involve sensations or feelings of strain from the exercising muscles and/or joints, and the central component has been associated primarily with cardiopulmonary sensations (Pandolf, 1977, 1978, 1982, 1986). Over the years, scientists have attempted to evaluate the relative importance of various physiological cues as they influence these local (peripheral) and central signals. Differentiated RPE have been used in many instances to help evaluate these local and central signals. In general, local or peripheral sensory signals have been reported to be perceived as dominant or primary while central signals act as amplifiers or gain modifiers that may potentiate the local signals (Cafarelli, 1977; Pandolf, 1978, 1982, 1986; Robertson, 1982). This scheme, which was developed in part from earlier concepts and observations by Cafarelli (1977), suggest that during the initial 30 s of exercise the exertional signals are totally local in origin. Central sensory signals are suggested to begin their potentiating effect after 30–180 s of exercise (Cafarelli, 1977; Robertson 1982).

Pandolf and Cain (1974) previously hypothesized that the rapidly declining portion of the CE function for cycling exercise primarily reflected the local component, while the slowly declining portion was associated primarily with the central component. If the sensory signals during the first few minutes of exercise are dominated by the local component, as suggested by both Cafarelli (1977) and Robertson (1982), the local component should also be dominant during the first few minutes of CE exercise, which supports the above hypothesis. Thus, local signals may be dominant in shaping the overall sense of effort during steady-state and CE exercise. We also suggest that some of the accountable variance in common between RPE and CE power output may primarily be associated with local and also possibly central signals.

In conclusion, these findings indicate that RPE and CE power output are related, particularly in those individuals displaying the hypothesized inverse relationship between these two variables. Local sensory signals seem to dominate during the first few minutes of both steady-state and CE exercise. When studied collectively, RPE and CE may someday help explain the mechanism(s) responsible for the buildup of fatigue during physical exercise. Further studies comparing RPE and CE appear warranted.

Acknowledgments. This paper was presented in part before the American College of Sports Medicine 34th Annual Meeting, Las Vegas, Nevada, May 1987.

The views, opinions and/or findings in this report are those of the authors, and should not be construed as an official Department of the Army position, policy, or decision, unless so designated by other official documentation. Human subjects participated in these studies after giving their free and informed voluntary consent. Investigators adhered to AR 70-25 and USAMRDC Regulation 70-25 on Use of Volunteers in Research. Approved for public release; distribution is unlimited.

References

Åstrand, I. (1960). Aerobic work capacity in men and women with special reference to age. *Acta Physiologica Scandinavica, 49* (Suppl. 169), 1–92.

Åstrand, P. O. & Rodahl, K. (1986). Evaluation of physical performance on the basis of tests. In M. Provenzano (Ed.), *Textbook of work physiology* (3rd ed., p. 356). New York: McGraw-Hill.

Borg, G. (1970). Perceived exertion as an indicator of somatic stress. *Scandinavian Journal of Rehabilitation Medicine, 2,* 92.

Cafarelli, E. (1977). Peripheral and central inputs to the effort sense during cycling exercise. *European Journal of Applied Physiology, 37,* 181–189.

Cain, W. S. (1973). Nature of perceived effort and fatigue: Roles of strength and blood flow in muscle contractions. *Journal of Motor Behavior, 5,* 33–47.

Cain, W. S. (1977). Versatility of the constant-effort procedure. In G. Borg (Ed.), *Physical work and effort* (pp. 49–60). Oxford: Pergamon.

Cain, W.S. & Stevens, J.C. (1971). Effort in sustained and phasic handgrip contractions. *American Journal of Psychology, 84,* 52–65.

Cain, W. S. & Stevens, J. C. (1973). Constant-effort contractions related to the electromyogram. *Medicine and Science in Sports, 5,* 121–127.

Kamon, E. & Pandolf, K. B. (1972) Maximal aerobic power during laddermill climbing, uphill running and cycling. *Journal of Applied Physiology, 32,* 467–473.

Pandolf, K. B. (1977). Psychological and physiological factors influencing perceived exertion. In G. Borg (Ed.), *Physical work and effort* (pp. 371–383). Oxford, Pergamon.

Pandolf, K. B. (1978). Influence of local and central factors in dominating rated perceived exertion during physical work. *Perceptual and Motor Skills, 46,* 683–698.

Pandolf, K. B. (1982). Differentiated ratings of perceived exertion during physical exercise. *Medicine and Science in Sports, 14,* 397–405.

Pandolf, K. B. (1986). Local and central factor contributions in the perception of effort during physical exercise. In G. A. V. Borg (Ed.), *The perception of exertion in physical exercise* (pp. 97–109). Hampshire, England: Macmillan (Wenner-Gren International series, Vol. 46).

Pandolf, K. B. & Cain, W. S. (1974). Constant effort during static and dynamic muscular exercise. *Journal of Motor Behavior, 6,* 101–110.

Pandolf, K. B., Billings, D. S., Drolet, L. L., Pimental, N. A., & Sawka, M. N. (1984). Differentiated ratings of perceived exertion and various physiological responses during prolonged upper and lower body exercise. *European Journal of Applied Physiology, 53,* 5–11.

Pandolf, K. B., Cafarelli, E., Noble, B. J., & Metz, K. F. (1972). Perceptual responses during prolonged work. *Perceptual and Motor Skills, 35,* 975–985.

Robertson, R. J. (1982). Central signals of perceived exertion during dynamic exercise. *Medicine and Science in Sports, 14,* 390–396.

Stevens, J. C. & Cain, W. S. (1970). Effort in isometric muscular contractions related to force level and duration. *Perception and Psychophysics, 8,* 240–244.

Stevens, J. C. & Krimsley, A. S. (1977). Buildup of fatigue in static work: Role of blood flow. In G. Borg (Ed.), *Physical work and effort* (pp. 145–155). Oxford: Pergamon.

Rating of Perception of Symptoms as an Aid in Medical Diagnostic Exercise Testing

HÅKAN LINDERHOLM

Abstract

Rating of perception of symptoms during diagnostic exercise tests and the development of its practical application are described. Rating of general exertion during exercise testing is an aid in conducting the test in patients. It also allows the determination of measures of exercise tolerance which sometimes have advantages compared with other methods. In healthy subjects the exercise tolerance according to rating of perceived exertion is closely related to other measures of exercise tolerance such as maximum work capacity and submaximal work capacity at given heart rates. Relationships between rating of exertion and physiological variables such as heart rate or work load are different for different types of work, e.g., arm and leg work; this should be taken into account when rating of exertion is used to evaluate, for instance, occupational work. In the standardized bicycle ergometer test these relationships also differ between groups of patients with some cardiovascular and pulmonary diseases.

When the exercise test is used to diagnose circulatory or respiratory disorders, rating exertion during the test improves its diagnostic accuracy. By relating the ratings to other variables recorded during the tests such as work load, heart rate, and respiratory frequency, variables are obtained which discriminate patients with different cardiopulmonary disorders. Rating of effort angina intensity during exercise in patients with coronary heart disease and relating this to other variables recorded during the test makes it possible to construct an index of the extent or severity of the coronary insufficiency. Its relationship to other indices of coronary heart disease is described. It may be concluded that rating of perceptions of different symptoms during an exercise test contributes valuable information and improves the diagnostic value of the test.

Introduction

A medical diagnosis is often founded on one or several symptoms or signs. For a qualitative diagnosis one specific symptom or sign may be enough, but more usually the diagnosis is founded on a combination of several symptoms or signs.

Many symptoms or signs appear only in situations of functional stress, e.g., during physical exercise. This has been used extensively in medical exercise testing used as an aid in diagnosis (Sjöstrand, 1960). The symptoms reported by a patient correspond to his or her perception of, for example, exertion, dyspnea, fatigue, and pain. In an exercise test it is, however, important that the symptoms be related to the type, intensity, and duration of the exercise as well as to other variables that can be recorded during the exercise test (Linderholm, 1986 a).

Exercise tolerance is more or less symptom limited. A low work capacity may in itself be a sign of disease. Rating of the intensity of the symptoms during the test may add important information and contribute to making the diagnosis more quantitative in nature, i.e., it may indicate the degree of deviation from the norm, and it may contribute in evaluating exercise tolerance.

I will give some examples of the way in which rating of symptoms during exercise may contribute to evaluating exercise tolerance and to medical diagnosis. I shall review some previously published and unpublished studies, confining the examples to ratings of general exertion and chest pain.

Historical Perspectives

Some historical perspectives on the development of psychophysical studies and particularly of studies on the perception of exertion in bicycle ergometry were recently given by Borg (1986). To this may be added some notes on the development of the use of rating of perceptions during bicycle exercise tests in medicine, and particularly its use for diagnostic purposes in Sweden.

After the introduction of a standardized bicycle ergometer exercise ECG test (Sjöstrand, 1947; Wahlund, 1948) for examination of patients, its use spread rapidly in Swedish hospitals with the establishment of departments of clinical physiology during the 1950s and 1960s. Borg and Dahlström (1962) described the use of rating of exertion in a patient (lumber worker) with atrial fibrillation to evaluate exercise tolerance. This patient's heart rate (HR) was high even at a low work load, indicating a low capacity for physical work. However, he had been able to perform heavy lumber work. The rating of exertion during bicycle ergometry indicated a normal capacity for physical work. This exemplified the value of the rating method for evaluation of exercise tolerance when there are disturbances of heart rhythm.

Later on, Borg and Linderholm (1967, 1970) examined the relationships between work load, rating of exertion, HR, and age in normal subjects and differences in these relationships in some patient groups (patients with coronary insufficiency and vasoregulatory asthenia). These early studies were followed by a number of others (Borg, Holmgren & Lindblad, 1981; Ekblom & Goldbarg, 1971; Sjöberg, Frankenheuser, & Bjurstedt, 1979), and the use of the rating of exertion during exercise spread. As the number of exercise tests performed in departments of clinical physiology in Sweden is high (in total about 50 000 per year), considerable experience of its use has accumulated. Rating of different symptoms during exercise such as anginal pain, muscular pain in the legs, and dyspnea has added further possible ways of evaluating the patients' condition. (Åström, 1986; Karlsson, Åström, Holmgren, Kaijser & Orinius, 1984; Killian, 1986; Pandolf, 1986). One advantage of using rating of exertion and other symptoms routinely is that it makes it easier for the person conducting the exercise test to decide when the work should be interrupted.

Nowadays the routine use of methods for rating of perceptions during exercise testing is well established in departments of clinical physiology in Sweden (Palm, Svensson & Wranne, 1987). When the significance of the ratings in the diagnostic evaluation of patients becomes more generally accepted, the importance of psychophysiological methods will probably be even more widely appreciated.

Methods

Exercise Test

The exercise test used in the experiments from which results reported in this paper were obtained was the bicycle ergometer test described by Sjöstrand (1947, 1960) and Wahlund (1948); it is used extensively in departments of clinical physiology in Sweden. The subject worked in a sitting position. The work loads may be chosen individually but usually the initial work load was 50 W for males and 30–40 W for females. The work load was increased every 6 min, usually by 50 W for males and 30 W for females, until the subject was unable to continue or symptoms such as pain or, for example, ECG signs made the tester stop the exercise. The test was thus essentially symptom limited.

In some experiments the bicycle ergometer was adapted for arm exercise (cranking) but the same test protocol was used.

Heart rate (HR) was measured from the recorded ECG at the end of each period at a particular work load and respiratory frequency (RF), arterial blood pressure (BP), and rating of perceived exertion (RPE) during the last minute at a particular work load or close to the end of the exercise for the highest work load.

Oxygen uptake, $\dot{V}O_2$, was measured using the Douglas' bag method, either at rest or during the last 1–2 min of exercise at each work load, or, for job tasks (see later), during periods with fairly stable HR and RPE.

Rating of Perceived Exertion

Different scales have been used for rating perceived exertion, and Borg has published several versions (see Ljunggren, 1985). The different rating scales may be more or less accurately transformed one into the other (Borg & Borg, 1987; Borg, Hassmén & Lagerström, 1987).

Table 1. The category scale (REA 6–20) used for rating of chest pain (effort angina) and corresponding figures on the category ratio scale (CR-10)

REA	Verbally anchored ratings (English and Swedish)	CR-10 scale
6		0
7	Extremely weak (Extremt Svag)	0.5
8		
9	Very weak (Mycket Svag)	1
10		
11	Weak (Svag)	2
12		
13	Moderate (Måttlig)	3
14		
15	Strong (Stark)	5
16		
17	Very strong (Mycket stark)	7
18		
19	Extremely strong (extremt stark)	10
20		

The CR-10 scale has lately been used extensively (Åström, 1986; Palm, Svensson, & Wranne, 1987). By means of an equation of a type suggested by Borg & Borg (1987) the REA 6–20 scale may be transformed into a CR-10 scale. The transformation equation $R_t = 1.95 \times 1.15^{(C-6)} - 1.95$ (where R_t is the transformed rating and C is the rating of the REA 6–20 scale) resulted in R_t values which were closely and linearly related to the verbally anchored CR-10 scale ratings ($r \times y = 0.9994$). This seems to justify and corroborate this type of transformation.

The category scale for rating perceived exertion that consists of 21 grades was used in some previous work (Borg & Linderholm, 1967, 1970), but in later work reported in this paper the RPE 6–20 scale was used (Linderholm, 1986a). The scales are appropriate for interindividual comparisons, for instance for diagnostic purposes, and may with minor modifications to the instructions also be used for rating of, for example, pain. The scale for rating chest pain in the experiments reported in this present paper is shown in Table 1.

Exercise Tolerance

The maximum work capacity (W_{max}) was defined as the maximum work load that the subject could perform for 6 min or, if the duration of exercise at the highest work load was less than 6 min, W_{max} was the highest work load reduced by the fraction of the latest stepwise increase in work load that corresponded to the fraction of the 6 min during which the subject was unable to perform the work (Strandell, 1964).

W_{max} has a wide distribution in normal subjects. A multiple regression equation, based on results obtained in a group of normal subjects, relating W_{max} and the independent variables age, height, weight, and sex was used by Linderholm, Osterman, and Teien (1985) to predict the normal maximum work capacity of a subject ($W_{max, N}$). By relating the determined W_{max} of a patient to the predicted normal value, the % W_{max} ($= 100 \times W_{max}/W_{max, N}$) was obtained; this was less variable in a normal group than W_{max}. % W_{max} is thus the actual work capacity of a subject as a percentage of the predicted normal value; it was used in the construction of some indices discussed below.

Rating of Perceived Exertion During Exercise and Evaluation of Exercise Tolerance

Bicycle Ergometry, Leg Work

Maximum work intensity (W_{max}) may be defined as a work load which cannot be increased further, or the corresponding oxygen uptake. In healthy subjects at this work load HR and the blood lactate concentration are high and the rating of exertion corresponds to verbal expressions such as very, very strenuous or extremely strenuous work. The maximum HR obtained varies between individuals and particularly with age, while the rated exertion at W_{max} is usually approximately the same.

In examining patients it is often desirable not to require maximum exertion. In such a case exercise tolerance may be estimated from submaximal work loads, e.g., from HR or rated exertion at work loads less than maximum. Such useful measures of submaximal exercise tolerance are physical work capacity at rating 13 or 17 of the RPE scale (WR13 or WR17) as well as the work capacity at a HR of 130 or 170 beats/min (W130 or W170). In normal subjects there are close relationships between these measures and also between them and W_{max}. These relationships are influenced by age (Borg & Linderholm, 1967, 1970). In patients these interrelationships may be different and such differences may have diagnostic significance (see below). Of interest in this context is that measures of submaximal exercise tolerance in a healthy subject may be estimated with comparable accuracy and precision by rating of exertion and by measuring HR at different work loads during the exercise test (Borg & Linderholm, 1970).

In physically active subjects (lumber workers) WR13 and WR17 seem to be fairly independent of age while W130 and W170 increase with age (Borg & Linderholm, 1967). In more "ordinary" groups of people WR13 and WR17 decrease more than W130 and W170 with increasing age. If W170 is corrected for the effect of age as in the PWC_p measure of exercise tolerance (Borg & Linderholm, 1970) it follows the variation of WR17 with age fairly well (Borg & Linderholm, 1967). WR17 also follows the variation of W_{max} with age fairly well and is a more constant fraction of W_{max} in individuals of different age than W170. It may be said that WR17 gives a more realistic picture of the exercise tolerance of a subject as a fraction of W_{max} than W170.

Exercise with Muscle Groups of Different Size

In bicycle ergometry, i.e., leg work with large muscle groups, the relationships between work load and RPE and between work load and HR show great interindividual variation, while the relationship between HR and RPE shows little interindividual variation in healthy subjects of the same age and sex (Borg & Linderholm, 1970). In another type of exercise, i.e., exercise with the arms (cranking on the bicycle ergometer), the relationship between RPE and HR is different, although interindividual variation is also small in this case (see Fig. 1). The submaximal work capacity of an individual with the arms is about 50%–60% of that with the legs. In work with even smaller muscle groups, e.g., hand work, HR and total metabolic rate increase very little while the work may be rated as very strenuous. Thus, the muscle mass engaged in the work greatly influences the psychophysiological and psychophysical relationships (see Linderholm, 1977; Borg et al., 1987). Minor discrepancies between the present results and those reported by Borg et al. (1987) may be due to different exercise test protocols.

Evaluation of Work Load and Physical Effort in Occupational Work (Job Tasks)

Rating of exertion or effort during work has been used with the intention to match the capacity of individuals to the physical requirements of jobs. The RPE value is a measure of the relative exertion of an individual in the job and may be related to a "zero point" of exertion at rest and an assumed rating of maximum exertion according to Borg's range theory (for references see Ljunggren, 1985). It may be compared with the traditional physiological measures of relative work load of an individual as expressed by the increase from resting condition in oxygen uptake ($\dot{V}O_2$) or HR when performing the job in question in relation to the corresponding increase in $\dot{V}O_2$ or HR at maximum work capacity (W_{max}).

The absolute work intensity of a job may be described in terms of $\dot{V}O_2$. It may also be obtained from an exercise test as the work load on the bicycle ergometer which results in the same HR or RPE as the job. However, this latter method of estimating the absolute work intensity of a job is only valid if the relationships between HR, RPE, and $\dot{V}O_2$ are the same in the job and in the bicycle ergometer test. This is usually not the case.

The differences that exist between different types of work may be exemplified by some experiments performed by medical students in some of my classes 1971–1973 when studying the physiology of occupational work. Aided by trained laboratory assistants, they measured HR and RPE in five women, 55 (45–61) years of age (mean and range). The tasks were (1) cleaning walls, (2) working in the hospital kitchen, and (3) working in the sterilization department at cleaning and packing, i.e., work that to a great extent was arm work.

Figure 2 shows the relationship between RPE and HR during the ergometer test and the job task. The rating of exertion at a given HR was higher

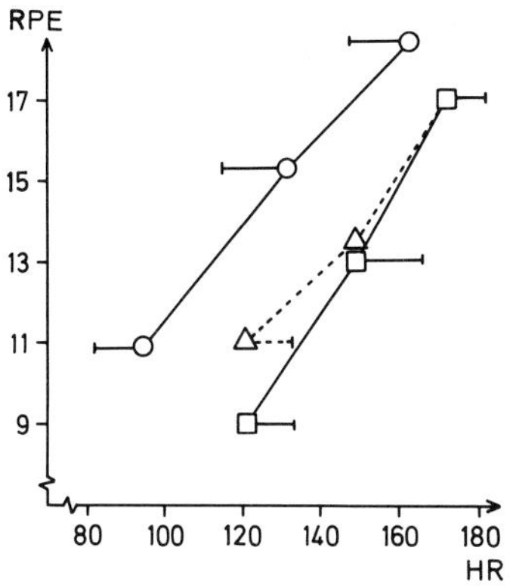

Fig. 1

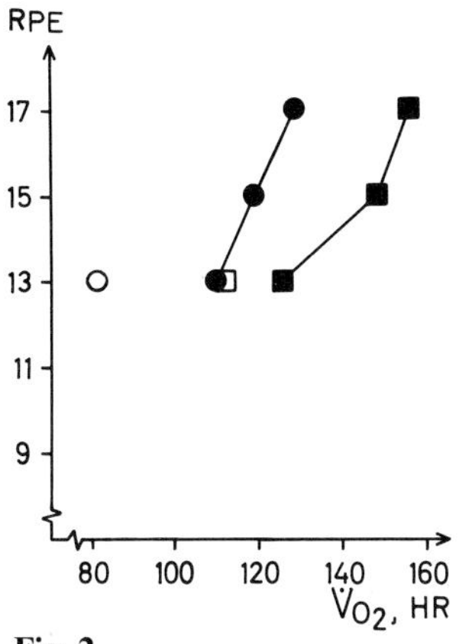

Fig. 2

Fig. 1. The relationship between rating of perceived exertion (RPE) and heart rate (HR) in arm (*circles*) and leg (*squares*) work on the bicycle ergometer. The *horizontal bars* indicate 1 SD of HR. Arm work results are from 24 healthy male subjects, mean age 24 years (range 19–39) (see Linderholm, 1977). Leg work results are from group N_I, 91 males and females, mean age 27 years (range 20–39). They are almost identical results to group N_{IV}, 24 males, mean age 23 years (range 20–24). Both these groups were described in a report by Borg & Linderholm (1970). In that investigation the 21-grade rating scale was used. It differs from the RPE 6–20 scale particularly in the low range of rating values. An approximate transformation of this scale to the RPE 6–20 scale is indicated by the *triangles*

Fig. 2. The relationship between RPE and HR during leg work on the bicycle ergometer (*filled squares*) and occupational work, mainly arm work (*filled circles*) for five women, mean age 55 years (range 45–61). The average HR at a given RPE in the two types of exercise is the mean of one value for each woman. For two of the women, aged 45 and 61 years, the mean oxygen uptake $\dot{V}O_2$ (100 ml (STPD)/min) at equal ratings, R13, is given for the bicycle (*open square*) and the occupa-tional work (*open circle*)

and the HR at a given rating was lower during the job task (mainly arm work) than during bicycle exercise with the legs. The difference was similar to that found between arm and leg cycling (Fig. 2). Because of the greater age of the subjects who did the job tasks, the RPE values at a given HR are in general higher than those of the young subjects whose results are shown in Fig. 1. In two of the women $\dot{V}O_2$ was also measured at equal ratings (R13) during the bicycle exercise test and during the job task. $\dot{V}O_2$ was, like HR, lower during the occupational work than during cycling (Fig. 2).

Thus, if $\dot{V}O_2$ is not measured, the absolute (and relative) work intensity of a job task which mainly engages arm and shoulder muscles can be estimated from an arm cranking test on the ergometer (up to W_{max}). When the job task involves long periods of work, a test protocol with stepwise increasing work loads and work periods of 6 min should be used to obtain a relative steady state. Then a fair estimate of the absolute (and relative) work intensity of the task may be obtained from recordings of RPE and/or HR.

Rating of Exertion During Exercise and its Contribution to the Diagnostic Accuracy of the Exercise Test

Differences in the relationship between HR and RF during a bicycle ergometer exercise test with stepwise increasing work load were used by Wahlund (1948) to differentiate patients with heart and lung disease. In the same way, differences in the relationship between HR and RPE during the exercise test may be used to differentiate groups of normal subjects, patients with coronary heart disease, and patients with vasoregulatory asthenia (Borg & Linderholm, 1970).

A more sophisticated method of differentiating between healthy subjects and patients with different diseases is to include more variables and use a multiple discriminant analysis (Linderholm, 1981, 1986a). In this way the bicycle ergometer exercise test may be used as a diagnostic tool for discriminating between patients with circulatory and pulmonary dysfunction.

During the exercise test the variables HR, RF, RPE, BP, and work load were recorded. From these, variables were derived. Some variables took account of deviations from the normal values obtained from examining healthy subjects.

The percentage deviation of exercise tolerance from the predicted normal value, % W_{max} D discriminated strongly between normal subjects and patients with circulatory and pulmonary disturbances, but the ability to discriminate between different groups of patients increases with increasing number of exercise test variables included in a multiple discriminant analysis. Of interest is that RPE during exercise contributes significantly to the discrimination between patients with cardiovascular and pulmonary diseases, not as a single rating value but as the rating of exertion in relation to RF (at R13) or HR (at R13 and R17; see Linderholm, 1986a).

Multiple discriminant analysis may also be used to differentiate between different types of pulmonary disorders. Karvonen and Linderholm (1981) described the use of an exercise test that included RPE to discriminate between obstructive, restrictive, and diffusion insufficiencies of the lungs. Also, in this case RPE in relation to RF and HR contributed significantly to the discrimination of these patient groups.

Groups consisting of healthy subjects and groups of patients with pulmonary obstructive, restrictive, and diffusion insufficiencies were used to construct a discriminant function. This was used to classify subjects of test groups selected according to the same criteria as the original groups. The discriminant function classified correctly the subjects of each of the test groups versus the other three groups with an overall diagnostic accuracy (DA) of 70%–90%. The discriminant function contained 11 discriminating variables (Table 2). It is of interest that five of the 11 variables included RPE, in relation to RF, work load, and HR. Rating of exertion thus contributed substantially to the diagnostic accuracy of the test.

It is also possible to discriminate between groups with different cardiovascular diseases. In this case variables derived from the ECG become more

Table 2. Exercise test variables which discriminate healthy subjects and patients with pulmonary obstructive, restrictive, and diffusion insufficiencies (from the work of Karvonen & Linderholm, 1981)

1 %WRF 25 D	The work load at a respiratory frequency of 25 breaths/min as the percentage deviation from the normal value
2 % W_{max} D	The maximum work capacity as the percentage deviation from the normal value (see Linderholm, Osterman, & Teien, 1985)
3 RF, R9	The respiratory frequency at rating 9
4 B (R, RF)	The slope of the regression line for the relation between rating (R) and respiratory frequency (RF)
5 %WR9 D	The work load at rating 9 as the percentage deviation from the normal value
6 A (E, RF)	Intercept (A) of the regression $RF = B \times E + A$; E is the work load.
7 %WR17	The percentage deviation from the normal value of the work load at rating 17
8 B: HR, R9	Slope of the straight line between two pairwise observations of heart rate (HR) and rating at or close to R9
9 DBP3 – DBP1	Differences between diastolic blood pressure during the orthostatic test (8 min standing) and at rest supine
10 %W150 AC D	The work load at an age-corrected (AC) heart rate corresponding to HR 150 at age 20 years as the percentage deviation from the normal value (Borg & Linderholm, 1967, 1970)
11 HR RF30	Heart rate at respiratory frequency 30 breaths/min

Normal values were obtained from results of examinations of healthy subjects using multiple regression equations with age, height, weight and sex as independent variables (see also List of symbols).

important, although the exercise test variables work load, HR, RF, BP, and RPE, and the variables derived from them still contribute in the discrimination between some groups.

Thus, in discriminating between groups of patients with (1) coronary insufficiency, (2) previous myocardial infarction, (3) cardiosclerosis (defined as a group with minor ECG changes in the exercise ECG but with a small increase in HR during the exercise test, a low HR in relation to rated exertion, and decreased exercise tolerance), (4) combined arterial hypertension and coronary insufficiency, and (5) previous myocardial infarction and coronary insufficiency, rating of exertion in relation to HR and work load [HR(R17), HR(R9), %WR13D] entered into the seven discriminant variables found when no ECG variables were used. When ECG variables were included in the discriminant analysis, HR(R17) still entered as the fourth discriminating variable. With another combination of diagnostic groups, including normal subjects and patients with arterial hypertension, myocarditis, and vasoregulatory asthenia, variables containing rating of exertion did not enter as discriminating variables (Linderholm, unpublished results).

Rating of Effort Angina Intensity During Exercise and Relationships with Other Indices of Coronary Heart Disease

In patients with coronary heart disease and effort angina, the intensity of angina usually has a typical time course during a standardized bicycle ergometer exercise test. Rating of the intensity of the angina during the test makes it easier to record its time course and follow the typical, more or less rapid, decline and disappearance of the angina after the termination of the exercise. Angina intensity is often closely related to the ST depressions of the ECG. The rating of effort angina (REA) as well as RPE helps the person conducting the exercise test to decide when the test should be interrupted but also has a diagnostic significance. Sometimes the typical ST depressions are absent, e.g., because of a left bundle branch block or a previous myocardial infarction, which may hide the presence of a coronary insufficiency. In such cases the typical time course of the effort angina intensity is of great diagnostic help.

In a group of patients with coronary heart disease ($n = 64$) who underwent a symptom-limited bicycle ergometer exercise ECG test that included REA, it was found that the REA at W_{max} was not much different in groups with greatly different decreases in W_{max} and greatly different angiocardiographic measures of the extent of the coronary heart disease (MCOS) or coronary insufficiency (MCOS – LVMS; see Table 3). The same was true for the ST depressions, which did not differ significantly between such groups (Linderholm 1986b).

Table 3. Average rating of effort angina (REA) and ST depression in two groups of patients with coronary heart disease, one with high (% $W_{max} > 50$) and the other with low (% $W_{max} < 35$) exercise tolerance. Mean % W_{max}, REA/W and ST/W indices, and angiocardiographic scores MCOS and MCOS – LVMS are also given

Exercise test variables	Patients with % $W_{max} > 50$ ($n = 16$)	Patients with % $W_{max} < 35$ ($n = 16$)	p
% W_{max}	63	25	< 0.001
REA	14 (4.0)	16 (5.9)	< 0.05
ST depr (10 × mV)	2.2	2.6	n.s.
REA/W index	2.3	7.3	< 0.001
ST/W index	3.4	10.8	< 0.001
MCOS	7.2	11.3	< 0.001
MCOS – LVMS	5.9	9.1	< 0.001

Within brackets after the means of REA are given the corresponding rating values according to the CR 0–10 scale. p is the probability that the differences between the two groups is caused by random factors. n.s., nonsignificant, according to the Mann-Whitney U test (Siegel, 1956).
The two groups consisted of the 16 patients with the lowest and the 16 with the highest % W_{max} out of a group of 64 patients who were evaluated consecutively for coronary bypass surgery and in whom effort angina was rated during the exercise test.

Table 4. Spearman rank correlation coefficients (Siegel, 1956) for the relationships between angiocardiographic scores of coronary artery disease (MCOS) and coronary insufficiency (MCOS – LVMS) on the one hand and REA, ST depr, and the REA/W and ST/W indices on the other in 64 patients with coronary heart disease

Exercise test variables	Angiocardiographic variables	
	MCOS	MCOS – LVMS
REA	0.32 [a]	0.35 [b]
ST depr	0.36 [b]	0.49 [c]
REA/W index	0.52 [c]	0.50 [c]
ST/W index	0.55 [c]	0.64 [c]

[a] $0.01 < p < 0.05$, [b] $0.01 < p < 0.001$, [c] $p < 0.001$

Effort angina and ST depr are regarded as indicating coronary insufficiency, i.e., myocardial ischemia, during exercise. However, it is not the angina intensity (REA) or the magnitude of the ST depressions at W_{max} which indicate best the extent or severity of the coronary insufficiency or myocardial ischaemia. Instead it is the decrease in exercise tolerance, i.e., the work intensity at which these comparatively large ST depressions and a high intensity of effort angina are present, which indicates the severity of the coronary insufficiency. Therefore, indices of the extent of coronary insufficiency were constructed by dividing 10 × REA or 100 × ST depression (in 10 × mV) by $\% W_{max}$ (see List of symbols). These indices (REA/W index and ST/W index) were more closely related to the angiocardiographic estimate of the extent of myocardium subserved by obstructed coronary arteries (MCOS), and particularly to the extent of coronary insufficiency indicated by MCOS – LVMS (see Backman, Billström, Eriksson, Jacobsson, Johnson, Linderholm, and Osterman, 1986), than were REA or ST depr at W_{max} (see Table 4).

A high REA and large ST depr should therefore be regarded more as symptoms and signs that a limit of exercise tolerance has been approached. It is interesting that this often occurs in patients with coronary heart disease when the RPE is still fairly low. An indication of the severity of the disease, i.e., a more quantitative diagnosis, is thus obtained when the REA or ST depr are related to another variable ($\% W_{max}$) to form the REA/W or ST/W indices.

Discussion and Conclusions

Rating of perceptions using different types of rating scales is extensively used in medicine, e.g., by medical psychologists. The development of rating of perceptions during medical exercise testing has been rapid and most likely we have only seen the beginning of its application. It obviously has potential to contribute both to the evaluation of exercise tolerance and also to the diagnostic accuracy of such tests.

RPE has added a dimension to the concept of exercise tolerance and rating of exertion is informative in patients with cardiac arrhythmias or conditions where HR changes abnormally with physical work. It has a value comparable to those of HR, blood lactate concentration, and RF in the evaluation of exercise tolerance. It has the advantage that it is easy to apply in field experiments and in evaluation of, for instance, occupational work (Fleishman, Gebhart & Hogan, 1986). It even seems possible to evaluate the exertion of patients during exercise by merely observing their behavior (Ljunggren, 1985), without knowing their subjective rating.

In different types of physical work different masses and groups of muscles are involved. This and the different types of exercise test protocols (Åström, 1986; Borg et al., 1981, 1987; Ekelund, Blumenthal, Morey & Ekelund, 1986; Sjöstrand, 1947, 1960) influence the psychophysiological and the psychophysical relationships. Further study of the physiological conditions which link the psychological and physical phenomena should be rewarding and elucidate factors which, when perceived and integrated by the individual, result in the quite accurate perception of exertion during physical exercise. By including ratings of dyspnea, pain in the chest and legs, etc. (Åström, 1986; Karlsson et al., 1984; Killian, 1986; Pandolf, 1986) further results of interest in medical diagnostics will probably be obtained.

It has been suggested that the rated exertion during work can be used to obtain information about physiological responses as well as about the physical effort involved (Fleishman et al., 1986). When different groups of people with job tasks involving a wide range of metabolic costs were examined, significant correlations were obtained between rated exertion and metabolic rate, but the variation was large (Fleishman et al., 1986). This is not astonishing when the results of the comparison between arm and leg work are considered.

In different types of work, e.g., arm and leg work, the relationships between the RPE and physiological variables are different. Therefore a close correlation between rated exertion and the physiological variables cannot be expected when different types of work are included in comparisons. Thus, rating of exertion during, for example, occupational work cannot be used to obtain accurate information about the physiological response in terms of metabolic cost or cardiovascular load, unless the relationships between rating of exertion and $\dot{V}O_2$, HR, and work load have been determined for the type of work in question. The size of the muscle mass engaged in the work is particularly important in this context.

Probably the rating of the exertion or effort required in a job or in different types of work gives the most direct and realistic estimate of the relative work load of the individual. However, in individual cases the rating values may be influenced by psychological factors in special situations and exertion may be over- or underrated. Therefore, despite it being easy to use in field as well as in laboratory experiments, the rating method often needs corroboration by means of the more objective physiological methods.

The exercise ECG test used in medical practice is quite rich in information. Even an experienced examiner may have difficulties in grasping this information and in the future multifactorial statistical methods will probably be used as aids in the evaluation. In this context, RPE will undoubtedly be one of the variables which should be included, and it should contribute to the diagnostic accuracy of the exercise ECG test (Karvonen & Linderholm, 1981; Linderholm, 1981, 1986a).

Promising for the use of rating in exercise testing is the development of ratings including not only general exertion but also various specified symptoms that may occur during exercise such as chest pain, dyspnea, and pains or exertion in the legs (Åström, 1986; Karlsson et al., 1984; Killian, 1986; Pandolf, 1986), and studies of the relationships of these ratings with other physiological variables recorded during the exercise test. From these, refined evaluations of circulatory and pulmonary diseases or functional circulatory disturbances may be obtained and the effect of therapeutic measures can also be evaluated (Ekelund et al., 1986).

A patient's rating of the perceived intensity of effort angina during and after an exercise test is of help in conducting the test. It shows the rate of increase in angina intensity with time and increase in work load. It also contributes information about the time course of the angina, which is valuable for the diagnostic evaluation of the test, particularly when ECG cannot be used for this purpose. When related to exercise performance it contributes to evaluation of the extent of coronary insufficiency.

List of Abbreviations

Angiocardiographic Scores

MCOS	Myocardial coronary obstruction score
LVMS	Left ventricular (wall motion) score
MCOS – LVMS	An estimate of the extent of coronary insufficiency (Backman et al., 1986)

Exercise Test Variables

BP	Arterial blood pressure
ECG	Electrocardiogram
HR	Heart rate (beats/min)
HR (R9), HR (R17)	Heart rate at R9, R17
PWC_p	See Borg and Linderholm (1967, 1970)
R11, R13, R17, etc.	Indicate RPE values 11, 13, 17, etc.
REA	Rating of effort angina according to the rating scale of Table 1
REA/W index	$10 \times \text{REA}/\% W_{max}$

RF	Respiratory frequency (breaths/min)
RPE	Rating of perceived exertion according to the RPE scale (see Linderholm, 1986a)
ST depr	Maximal increase in ST depression 0.04 or 0.06 s after the J-point (in 10 × mV) during or up to 4 min after exercise in any of the 12 standard extremity and chest leads of the ECG
ST/W index	100 × ST depr./% W_{max}
$\dot{V}O_2$	Oxygen uptake
W_{max}	Maximum work capacity (W)
% W_{max}	100 × $W_{max}/W_{max, N}$ where $W_{max, N}$ is the normal W_{max} predicted from age, sex, height, and weight (see Linderholm et al., 1985)
W130, W170	Work load at HR130 and 170
WR13, WR17	Work load at R13 and R17
% WR13, etc.	The work load at R13, etc., as a percentage of the normal value

Other Symbols

DA	Overall diagnostic accuracy DA = (TP + TN)/N, where TP is number of true-positive tests, TN is number of true-negative tests, and N is the total number of subjects examined
D	Added to other symbols indicates the percentage deviation from the normal values

References

Åström, H. (1986). Rating perceived symptoms during exercise in heart disease. In G. Borg & D. Ottoson (Eds.), *The perception of exertion in physical work. Proceedings of an international symposium held at the Wenner-Gren Center in Stockholm, Oct 3–5, 1985* (pp. 199–206). London: Macmillan (Wenner-Gren International Symposum series, Vol. 46).

Backman, C., Billström, A., Eriksson, M., Jacobsson, K.A., Johnson, O., Linderholm, H., & Osterman, G. (1986). Relationships between coronary artery obstruction, asynergy, presence of collaterals and the ejection fraction of the left ventricle in patients with coronary heart disease. *Acta Radiologica: Diagnosis, 27,* 413–418.

Borg, G. (1986). Psychophysical studies of effort and exertion: Some historical, theoretical and empirical aspects. In G. Borg & D. Ottoson (Eds.), *The perception of exertion in physical work. Proceedings of an international symposium held at the Wenner-Gren Center in Stockholm, Oct 3–5, 1985* (pp. 3–12). London: Macmillan (Wenner-Gren International Symposium series, Vol. 46).

Borg, G. & Borg, P. (1987). *On the relations between category scales and ratio scales and a method for scale transformation* (Report No. 672). Stockholm: University of Stockholm, Department of Psychology.

Borg, G. & Dahlström, H. (1962). Case study of perceived exertion during a work test. *Acta Societatis Medicorum Upsaliensis 67,* 91–93.

Borg, G. & Linderholm, H. (1967). Perceived exertion and pulse rate during graded exercise in various age groups. *Acta Medica Scandinavia Supplementum, 472,* 194–204.

Borg, G. & Linderholm, H. (1970). Exercise performance and perceived exertion in patients with coronary insufficiency, arterial hypertension and vasoregulatory asthenia. *Acta Medica Scandinavica, 187,* 17–26.

Borg, G., Hassmén P., & Lagerström, M. (1987). Perceived exertion related to heart rate and blood lactate during arm and leg exercise. *European Journal of Applied Physiology, 65,* 679–685.

Borg, G., Holmgren, A., & Lindblad, I. (1981). Qualitative evaluation of chest pain. *Acta Medica Scandinavica Supplementum, 644,* 43–45.

Ekblom, B. & Goldbarg, A. N. (1971). The influence of physical training and other factors on the subjective rating of perceived exertion. *Acta Physiologica Scandinavica, 83,* 399–406.

Ekelund, L. G., Blumenthal, J. A., Morey, M. C., & Ekelund, C. C. (1986). The effect of nonselective and selective beta blockade on perceived exertion during treadmill exercise in mild hypertensive type A and B males and the interaction with airobic training. In G. Borg & D. Ottoson (Eds.), *The perception of exertion in physical work. Proceedings of an international symposium held at the Wenner-Gren Center in Stockholm, Oct 3–5, 1985* (pp. 191–198). London: Macmillan (Wenner Gren International Symposium series, Vol. 46).

Fleishman, E. A., Gebhardt, D. L., & Hogan, J. C. (1986). The perception of physical effort in job tasks. In G. Borg & D. Ottoson (Eds.), *The perception of exertion in physical work. Proceedings of an international symposium held at the Wenner-Gren Center in Stockholm, Oct 3–5, 1985* (pp. 225–242). London: Macmillan (Wenner-Gren International Symposium series, Vol. 46).

Karlsson, J. Åström, H., Holmgren, A. Kajiser, L., & Orinius, E. (1984) Angina pectoris and blood lactate concentration during graded exercise. *International Journal of Sports Medicine, 5,* 348–351.

Karvonen, J. & Linderholm, H. (1981). Computerized discrimination between different types of pulmonary dysfunction by means of exercise testing. *Clinical Physiology, 7,* 594–595.

Killian, K. J. (1986). Breathlessness – The sense of respiratory muscle effort. In G. Borg & D. Ottoson (Eds.), *The perception of exertion in physical work. Proceedings of an international symposium held at the Wenner-Gren Center in Stockholm, Oct 3–5, 1985* (pp. 71–82). London: Macmillan (Wenner-Gren International Symposium series, Vol. 46).

Linderholm, H. (1977). Discussion of "Psychophysiological interactions leading to increased effort, leg fatigue, and respiratory distress during prolonged strenuous bicycle riding" by P. C. Weiser and D. A. Stamper. In G. Borg (Ed.), *Physical work and effort. Proceedings of the first international symposium held at the Wenner-Gren Center, Stockholm, December 2–4, 1975* (p. 415). Oxford: Pergamon.

Linderholm, H. (1981). Computerized discrimination between patients with circulatory and pulmonary dysfunctions by means of an exercise test. In *Computers in cardiology* (pp. 137–140). Long Beach, California: Institute of Electrical and Electronics Engineers, Computer Society.

Linderholm, H. (1986a). Perceived exertion during exercise in the discrimination between circulatory and pulmonary disorders. In G. Borg & D. Ottoson (Eds.), The perception of exertion in physical work. *Proceedings of an international symposium held at the Wenner-Gren Center in Stockholm, Oct 3–5, 1985* (pp. 177–190). London: Macmillan (Wenner-Gren International Symposium Series, Vol. 46).

Linderholm, H. (1986b). Relationships between severity of effort angina at an exercise test and coronary artery obstructive disease. *Clinical physiology, 6,* 92–93.

Linderholm, H., Osterman, G., & Teien, D. (1985). Detection of coronary artery disease by means of exercise ECG in patients with aortic stenosis. *Acta Medica Scandinavica, 218,* 181–188.

Ljunggren, G. (1985). *Studies of perceived exertion during bicycle ergometer exercise – Some applications*. Thesis, University of Stockholm, Department of Psychology.

Palm, O., Svensson, S.-E., & Wranne, B. (Eds.). (1987). *Det kliniska arbetsprovet* [The clinical exercise test]. Lund, Sweden: Sigma.

Pandolf, K. B. (1986). Local and central factor contributions in the perception of effort during physical exercise. In G. Borg & D. Ottoson (Eds.), *The perception of exertion in physical work. Proceedings of an international symposium held at the Wenner-Gren Center in Stockholm, Oct 3–5, 1985* (pp. 97–109). London: Macmillan (Wenner-Gren International Symposium series, Vol. 46).

Siegel, S. (1956). *Nonparametric statistics for the behavioral sciences*. New York: McGraw-Hill.

Sjöberg, H., Frankenheuser, M., & Bjurstedt, H. (1979). Interactions between heart rate, psychomotor performance and perceived effort during physical work as influenced by beta adrenergic blockade. *Biological Physiology, 8*, 31–43.

Sjöstrand, T. (1947). Changes in the respiratory organs of workmen at ore smelting works. *Acta Medica Scandinavica Supplementum, 196*, 687–699.

Sjöstrand, T. (1960). Functional capacity and exercise tolerance in patients with impaired cardiovascular function. In B. L. Gordon (Ed.), *Clinical cardiopulmonary physiology* (pp. 201–219). New York: Grune and Stratton.

Strandell, T. (1964). Circulatory studies on healthy old men. *Acta Medica Scandinavica Supplementum, 414*.

Wahlund, H. (1948). Determination of physical working capacity. *Acta Medica Scandinavica Supplementum, 215*.

Differential Effect of Blood Lactic Acid and pH on Peripheral Exertional Perceptions *

Robert J. Robertson and Kenneth F. Metz

Abstract

Induced alkalosis was used to determine the differential effect of blood lactic acid concentration ([HLa]) and pH on peripheral perceptions of exertion during arm and leg exercise. Ten males underwent exercise testing in three modes, arms (A), legs (L), and combined arms and legs (A + L), after ingesting 0.3 g kg^{-1} body weight of either $NaHCO_3$ to induce alkalosis or $CaCO_3$ as a placebo (i.e., six exercise trials). Ratings of perceived exertion (RPE; Borg 15-point scale) were determined for the arms (RPE-A) and legs (RPE-L). At 80% of maximum oxygen consumption ($\dot{V}O_2$ max): (1) RPE-A was lower ($p < .01$) during A; (2) RPE-L was lower ($p < .01$) during L; and (3) RPE-A and PRE-L were lower ($p < .01$) during A + L in the $NaHCO_3$ than in the $CaCO_3$ condition (Robertson, Falkel, Drash, Swank, Metz, Spungen & LeBoeuf, 1986). Differences between acid-base conditions were not found for RPE-L during A or RPE-A during L. Blood [HLa] and pH (Robertson et al., 1986) were not differentially influenced by exercise mode. Both blood [HLa] and pH were higher ($p < .01$) in the $NaHCO_3$ than in the $CaCO_3$ condition at 80% $\dot{V}O_2$ max. RPE-A and RPE-L were inversely related to [HLa]. During induced alkalosis, rated perceptions of exertion for the active limbs were mediated by blood pH but not by blood [HLa].

Introduction

During high-intensity dynamic exercise, blood pH and lactic acid concentration ([HLa]) are thought to mediate peripheral perceptions of exertion in active limbs (Allen & Pandolf, 1977; Kostka & Cafarelli, 1982; Noble, Borg, Jacobs, Ceci, & Kaiser, 1983; Young, Cymerman, & Pandolf, 1982). While the role of blood pH as a physiological mediator has recently been demonstrated (Kostka & Cafarelli, 1982; Robertson et al., 1986), the influence of [HLa] in mediating perceptions of exertion and fatigue is equivocal. Substantial correlational evidence suggests a link between [HLa] and rated perceptions of exertion (RPE)

---

* This investigation was supported in part by the University of Pittsburgh's Education Alumni Fund

for the body overall and for differentiated responses in the arms and legs (Pandolf, 1983). In contrast, experimental paradigms that have manipulated [HLa] to determine if exertional perceptions exhibit corresponding changes have not consistently demonstrated a causal relationship between variables (Pandolf, 1983). It is possible that blood pH shifts are more important than [HLa] in mediating peripheral perceptions of exertion. The purpose of this investigation was to examine this hypothesis.

The present investigation employed a research paradigm wherein acid-base perturbation during high-intensity exercise was accomplished by $NaHCO_3$ ingestion. Robertson et al. (1986) recently demonstrated that at an exercise intensity equivalent to $\cong 80\%$ of maximum oxygen consumption ($\dot{V}O_2$ max), blood pH was higher and peripheral ratings of perceived exertion lower following ingestion of $NaHCO_3$ (i.e., induced alkalosis) than after $CaCO_3$ (i.e., placebo). Blood [HLa] during exercise is also higher with alkalosis than in placebo conditions (Jones, Sutton, Taylor, & Toews, 1977; Sutton, Jones, & Toews, 1981). As such, perceptions of exertion arising from peripheral skeletal muscle should vary inversely with blood [HLa] following $NaHCO_3$ ingestion. It was anticipated that under conditions of induced alkalosis, rated perceptions of exertion for the active limbs would be mediated by blood pH and not by blood [HLa].

Differentiated ratings of perceived exertion were determined for active and inactive limbs during arm, leg, and combined arm and leg exercise (Robertson et al., 1986). By varying limb involvement between different exercise modes, it was possible more directly to link blood [HLa] and pH with differentiated peripheral perceptions arising from active muscle mass (Robertson, Gillespie, McCarthy & Rose, 1979).

Methods

Subjects

Ten physically active male university students participated as subjects. On average, they were 23.2 (SD $\pm$3.0) years old, 174 (SD $\pm$2.5) cm tall, weighed 73.2 (SD $\pm$7.0) kg, and had 7.7 (SD $+$0.8)% body fat and a cycle ergometer $\dot{V}O_2$ max of 3.50 (SD $\pm$0.47) $l\,min^{-1}$. Subjects were informed of the possible side effects of the experiment and gave their written consent to participate.

Experimental Design

A counterbalanced, double-blind design was used with each subject randomly assigned to one of six combinations of experimental trials. In this design, separate exercise tests were administered for the arms (A), legs (L), and combined arms and legs (A + L). Each of the three exercise modes was performed under two experimental conditions: (1) after ingesting $NaHCO_3$ to induce alkalosis, and (2) after ingesting $CaCO_3$, which served as the placebo control. A 7-day period separated each of the six trials. The dose for both the $NaHCO_3$ and $CaCO_3$ conditions was 0.3 g kg^{-1} body weight; this was consumed with 500 cm^3 of water over a 15-min period 2 h prior to testing. A complete description of the procedures used in this investigation has been published elsewhere (Robertson et al., 1986).

Exercise Tests

All exercise tests were administered on a specially adapted arm and leg ergometer consisting of a Monark 864 cycle ergometer and a Monark 881 Rehab Trainer. The Rehab Trainer was modified with a weight platform identical to that used on the Model 864 leg ergometer. All tests were administered with the subject's head, chest, and hips harnessed (5.1 cm Velcro webbing) to a stabilizing backboard which was permanently attached to a wall (Davies & Sargeant, 1974). By stabilizing the body, involvement of the trunk muscles was minimized, providing more complete isolation of the limbs during exercise (Davies & Sargeant, 1974).

Power output (watts) equivalents of 20%, 40%, 60%, and 80% of $\dot{V}O_2$ max were determined separately for A and L exercise. These power outputs were then used for the A and L exercise tests in the experimental trials. The A + L exercise test used the power outputs for the separate A and L tests. During the six exercise tests, each power output stage was 3 min. The four power outputs were presented progressively in ascending order.

Perceived Exertion

Ratings of perceived exertion (RPE) were measured using the 15-point Borg (1970) category scale. A differentiated rating was assessed for localized feelings of exertion in the arms (RPE-A) and for the legs (RPE-L). Prior to each exercise trial, a tape-recorded definition of the perception of exertion and instructions for use of the Borg RPE scale were played (transcript available from authors). The instructions emphasized that the perceptual ratings should reflect sensations of exertion, stress, and/or discomfort in the limbs. During testing, the perceptual ratings were measured from 15 to 45 s of each exercise minute. The rating scale was in full view of the subject during the entire test. As the respiratory valve prohibited a verbal rating response, a finger signal was employed (Robertson, Caspersen, Allison, Skrinar, Abbott, & Metz, 1982). Each rating was limited to a single numbered response.

Blood Analysis

Blood samples were obtained from an antecubital vein immediately following the 80% $\dot{V}O_2$ max exercise level. Whole blood pH was measured using a Corning 165 Blood Gas Analyzer. Blood samples for plasma [HLa] (mmol l^{-1}) determination were placed in 6% perchloric acid solution, agitated, and centrifuged. The supernatant was then frozen. Enzymatic analysis of [HLa] was performed according to Bergmeyer (1974, pp. 1463–1468).

Statistical Analysis

Perceived exertion responses were statistically analyzed by a three-factor (i.e., RPE by acid-base condition by exercise intensity) analysis of variance (ANOVA) with repeated measures on the first two factors (BMDP2V, Computerized Statistical Package). A two-factor repeated measures ANOVA was used to analyze blood pH and [HLa] (i.e., sampling period by acid-base condition). Separate analyses were undertaken for each exercise mode.

The relationship between RPE and [HLa] was determined by linear regression (University of Pittsburgh Interpretive Language).

Results

Differentiated Perceptual Responses

Means and standard deviations for perceived exertion responses in both acid-base conditions during A, L, and A + L exercise at 80% of $\dot{V}O_2$ max are listed in Table 1. (Note: These data have been published previously; Robertson et al., 1986). The effect of induced alkalosis on differentiated RPEs varied with the exercise mode. During A exercise, RPE-A was lower ($p < .01$) for the $NaHCO_3$ condition, while RPE-L did not differ between acid-base conditions. During L exercise, RPE-L was lower ($p < .01$) under alkalosis, while RPE-A did not change with acid-base shifts. For combined A + L exercise, RPE-A and RPE-L were both lower ($p < .01$) in the $NaHCO_3$ than in the $CaCO_3$ conditions.

Table 1. Effect of induced alkalosis on differentiated ratings of perceived exertion for the arms (RPE-A) and legs (RPE-L) during separate and combined arm and leg exercise (Robertson, Falkel, Drash, Swank, Metz, Spungen, & LeBoeuf, 1986)

RPE (15-point scale)	Condition	Exercise mode					
		Arms		Legs		Arms and legs	
		Mean	SD	Mean	SD	Mean	SD
RPE-A	$NaHCO_3$	14.9[a]	1.1	7.9	0.7	15.1[a]	1.5
	$CaCO_3$	16.8	0.9	7.9	0.9	17.4	1.0
RPE-L	$NaHCO_3$	7.9	1.0	14.6[a]	1.0	14.8[a]	1.0
	$CaCO_3$	7.8	0.9	16.6	1.0	16.9	1.2

[a] For each differentiated RPE, means are significantly different ($p < 0.01$) between acid-base conditions within a given mode

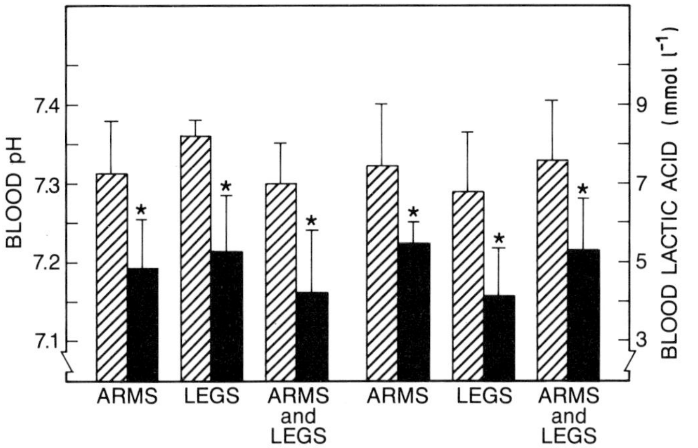

Fig. 1. Effect of induced alkalosis on venous blood pH (*left*) (Robertson, Falkel, Drash, Swank, Metz, Spungen, & LeBoeuf, 1986) and lactic acid concentration (*right*) during arm and leg exercise at 80% of $\dot{V}O_2$ max. For each exercise mode, an *asterisk* indicates that means (*bars* indicate standard deviations) are significantly different ($p < .01$) between acid-based conditions (*hatched bars,* with $NaHCO_3$; *filled bars,* with $CaCO_3$)

Lactic Acid and pH Responses

Means and standard deviations for venous blood [HLa] and pH for the three exercise modes are presented in Fig. 1. Both [HLa] and pH were higher ($p < .01$) in the $NaHCO_3$ than in the $CaCO_3$ condition during A, L, and A + L exercise. (Note: Venous blood pH data have been published previously as H + concentrations; Robertson et al., 1986.)

The relationships between peripheral ratings of exertion and blood [HLa] are shown in Fig. 2. Regression analysis of RPE-A as a function of [HLa] during A exercise and RPE-L as a function of [HLa] during L exercise are presented in parts a and b, respectively. The relationships between blood [HLa] and both RPE-A and RPE-L during A + L exercise are presented in parts c and d. Each regression analysis includes individual data points from the $NaHCO_3$ and $CaCO_3$ conditions.

Cardiorespiratory Responses

$\dot{V}O_2$ and heart rate responses did not differ between acid-base conditions for any mode. These data are presented elsewhere (Robertson et al., 1986).

Discussion

Metabolic acidosis is thought to mediate the intensity of exertional perceptions in peripheral skeletal muscle during high-intensity exercise (Robertson et al.,

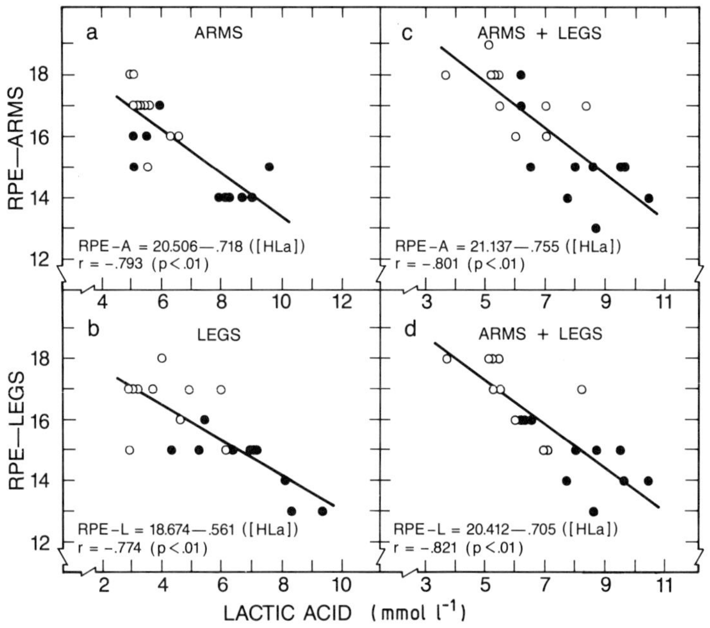

Fig. 2a–d. Regression analysis of the relationship between blood [HLa] and differentiated ratings of perceived exertion (RPE) during separate and combined arm and leg exercise at 80% of $\dot{V}O_2$ max. *Filled symbols*, with $NaHCO_3$; *open symbols*, with $CaCO_3$

1986). It was anticipated that with induced alkalosis, differentiated perceptions of exertion in active limbs would be mediated by blood pH shifts but would be independent of blood [HLa]. The findings confirmed this hypothesis. Rated exertional perceptions in the active limbs were lower in the $NaHCO_3$ than in the $CaCO_3$ condition in all three exercise modes. Blood pH was higher in the alkalotic than in the placebo condition, suggesting that shifts in acid-base balance mediate the intensity of peripheral exertional perceptions (Robertson et al., 1986). In contrast, blood [HLa] was higher following $NaHCO_3$ than following $CaCO_3$ ingestion, calling its role as a physiological mediator for perceptions of exertion in peripheral muscle mass into question.

Blood Lactic Acid: Absence of a Mediating Effect

Considerable experimental evidence both supports (Ekblom & Goldbarg, 1971; Gamberale, 1972; Noble et al., 1983; Young et al., 1982) and rejects (Allen & Pandolf, 1977; Lollgen, Graham, & Sjogaard, 1980; Robertson et al., 1979; Sargeant & Davies, 1973) blood [HLa] as a physiological mediator for exertional perceptions. These conflicting findings can be explained by separating the sensory effects of blood [HLa] from those of blood pH. Alkalosis induced via $NaHCO_3$ ingestion provides a mechanism to perturb both blood

[HLa] and pH during high-intensity exercise. As both [HLa] and pH are higher with alkalosis than in placebo conditions (Jones et al., 1977; Sutton et al. 1981) their separate effects on peripheral exertional perceptions can be studied simultaneously. That is, the intensity of exertional perceptions in active limbs should vary directly with blood [HLa] and inversely with blood pH, assuming both variables mediate the effort sense. However, with induced alkalosis the normal inverse relationship between blood pH and [HLa] is altered. The relatively more alkaline environment following $NaHCO_3$ ingestion prolongs glycolytic production of ATP for muscle contraction, resulting in more complete use of muscle glycogen stores and greater production of HLa (Jones et al., 1977; Kostka & Cafarelli, 1982; Sutton et al., 1981). As a result, [HLa] is inversely related to peripheral exertional perceptions (Fig. 2) arising from active limbs during A, L, and A + L exercise. The reciprocal of these findings has been reported by Kostka and Cafarelli (1982), i.e., exertional sensations were higher and blood [HLa] lower with induced acidosis (NH_4Cl ingestion) than in a placebo conditon. Therefore, during high-intensity dynamic exercise where blood acid-base status is perturbed by induced alkalosis or acidosis, [HLa] is not found to effect the intensity of exertional perceptions in peripheral skeletal muscle.

What, if any, is the importance of blood [HLa] in studying physiological correlates of perceived exertion? A partial answer to this question is found in the strong, positive relationship between power output and both rated perceptions of exertion and blood [HLa] during dynamic exercise. Borg, Ljunggren, & Ceci (1985), Borg, Hassmén, & Lagerström (1987), and Noble et al. (1983) report that RPE and [HLa] vary as a curvilinear function of power output when cycle ergometer exercise intensity is progressively incremented. It follows that RPE and blood [HLa] are also related under progressive exercise conditions (Borg et al., 1985, 1987). However, such a relationship does not necessarily imply causality. Rather, both RPE and blood [HLa] serve as specific cases of a more generalized psychophysiological response which indicates the relative intensity of exercise (Borg et al., 1987).

The correspondence between the appearance of HLa in blood and perceptual responsiveness appears to hold at higher (Noble et al., 1983) but not lower (Johansson, 1986) exercise intensities owing to limited anaerobic glycolysis at lower metabolic rates. This suggests that the interrelationship between power output, blood [HLa], and RPE may not take on practical value as a means of evaluating exercise intensity until a metabolic level such as the lactate inflection point is exceeded (De Mello, Cureton, Boineau & Singh, 1987; Robertson et al., 1986). The application of RPE in this manner can be undertaken without assuming a causal link between the appearance of HLa in blood and peripheral perceptual responsiveness. Each variable is an independent but corresponding marker of exercise intensity.

Blood pH: Mediating Effect

Differentiated perceptions of exertion arising from active limbs during A, L, and A + L exercise were attenuated following alkalotic shifts in blood pH[1]. Kostka and Cafarelli (1982) demonstrated that when acidotic shifts in blood pH were induced by NH_4Cl ingestion, exertional sensations were intensified. The reciprocal perceptual responses between these two experiments add support to the idea that blood pH has a role as a physiological mediator for peripheral perceptions of exertion. However, as Robertson et al. (1986) observed, blood pH mediates peripheral exertional perceptions at higher but not lower metabolic rates. Possibly acid-base status does not begin to mediate peripheral exertional signals until a metabolic marker such as the lactate inflection point is reached (Robertson et al., 1986).

The mechanism underlying blood pH as a mediator for peripheral perceptions of exertion involves a disruption of contractile and energy-producing properties of skeletal muscle (Kostka & Cafarelli, 1982; Poulus, Docter, & Westra, 1974; Robertson et al., 1986). The relative acidosis that accompanies high-intensity exercise interrupts glycolytic production of ATP for myofibril contraction, causing fatigue of metabolically active muscles (Jones et al., 1977; Sutton et al., 1981). As muscles fatigue, motor unit recruitment and firing frequency are increased (Cafarelli, 1977; Kostka & Cafarelli, 1982; Lollgen et al., 1980). The final common neurophysiological pathway for peripheral signals of exertion is directly linked to these alterations in motor unit activation. This pathway follows a feedforward loop wherein the number of central motor outflow commands required to maintain contractile force is increased as muscles fatigue. Corollary discharges carrying a copy of the descending commands are simultaneously transmitted to the sensory cortex (McCloskey, Gandevia, Potter, & Colebatch, 1983). The magnitude of these corollary discharges determines the intensity of the peripheral signal of exertion arising from active skeletal muscle.

The deleterious effect of metabolic acidosis on muscle contractile properties is comparatively less pronounced with induced alkalosis, as is the magnitude of central motor outflow commands. The sensory signals influenced by these neurophysiological events are comparatively less intense (Robertson et al., 1986).

Conclusions

The assessment of differentiated ratings of exertion for the three exercise modes under conditions of induced alkalosis made it possible to examine the differential effects of blood [HLa] and pH in mediating peripheral exertional perceptions. The findings indicate that blood [HLa] is not a physiological

---

[1] While these findings were reported previously (Robertson, Falkel, Drash, Swank, Metz, Spungen, & LeBouef, 1986), they were derived simultaneously with the blood [HLa] responses described presently.

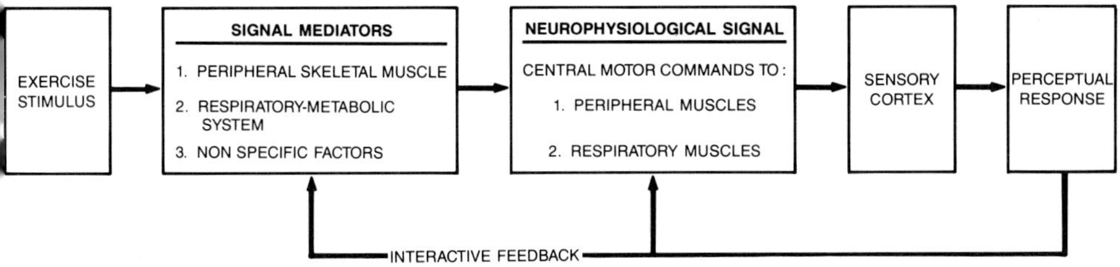

Fig. 3. Classification system for physiological mediators and neurophysiological signals of perceived exertion during dynamic exercise. (Modified from Robertson & Metz, 1986)

mediator for peripheral signals of exertional discomfort and fatigue arising from active limbs during high-intensity exercise. In contrast, shifts in blood pH appear to mediate peripheral exertional signals, particularly at metabolic rates in excess of the lactate inflection point.

The results of this experiment reinforce the premise originally advanced by Borg (1962) that exertional perceptions during dynamic exercise are linked to discrete underlying physiological events. Figure 3 presents a model of the sensory processes by which a discrete physiological event such as a shift in blood pH mediates the intensity of exertional perceptions (Robertson & Metz, 1986). The model is interpreted by moving from left to right. Physiological responses to an exercise stimulus serve as mediators for perceptual signals of exertion. These signal mediators act individually or collectively to alter tension-producing properties of skeletal muscle. In turn, change in peripheral and respiratory muscle tension are linked to the final common neurophysiological pathway for signals of exertion. It is this neurophysiological signal that is consciously interpreted by the sensory cortex as effort sensation.

An important feature of this psychophysiological model concerns the classification of physiological events that mediate perceptual signals. The classification system presented here represents a partial departure from that which is commonly used. *Peripheral physiological mediators* are localized to the limbs and trunk. *Repiratory-metabolic mediators*[2] comprise a cluster of physiological responses that influence ventilatory drive during exercise. A third class, termed *nonspecific mediators,* comprises generalized or systemic physiological events which are not directly linked to either peripheral or respiratory-metabolic signals. Nevertheless, these nonspecific mediators do contribute to effort sensations during exercise.

---

[2] The term *respiratory-metabolic* is used here in place of the more widely accepted term *central* when describing physiological mediators for the effort sense. This change in terminology is suggested for two reasons. First, the term respiratory-metabolic is connotatively consistent with those physiological events that influence breathing. Secondly, the term central normally refers to the central nervous system and thus its use is confusing when applied to respiratory-metabolic processes.

References

Allen, P. & Pandolf, K. (1977). Perceived exertion associated with breathing hyperoxic mixtures during submaximal work. *Medicine and Science in Sports, 9,* 122–127.

Bergmeyer, H. (1974). *Methods of enzymatic analysis,* Vol. 3. New York: Academic.

Borg, G. (1962). *Physical performance and perceived exertion.* Lund, Sweden: Gleerup.

Borg, G. (1970). Perceived exertion as an indicator of somatic stress. *Scandinavian Journal of Rehabilitation Medicine, 2,* 92–98.

Borg, G., Ljunggren, G., & Ceci, R. (1985). The increase of perceived exertion, aches and pain in the legs, heart rate and blood lactate during exercise on a bicycle ergometer. *European Journal of Applied Physiology, 54,* 343–349.

Borg, G., Hassmén, P., & Lagerström, M. (1987). Perceived exertion related to heart rate and blood lactate during arm and leg exercise. *European Journal of Applied Physiology, 65,* 679–685.

Cafarelli, E. (1977). Peripheral and central inputs to the effort sense during cycling exercise. *European Journal of Applied Physiology, 37,* 181–189.

Davies, C. & Sargeant, A. (1974). Physiological responses to standardized arm work. *Ergonomics, 17* (1), 41–49.

DeMello, J., Cureton, K., Boineau, R., & Singh, M. (1987). Ratings of perceived exertion at the lactate threshold in trained and untrained men and women. *Medicine and Science in Sports and Exercise, 19* (4), 354–362.

Ekblom, B. & Goldbarg, A. (1971). The influence of training and other factors on the subjective rating of perceived exertion. *Acta Physiologica Scandinavica, 83,* 399–406.

Gamberale, F. (1972). Perceived exertion, heart rate, oxygen uptake and blood lactate in different work operations. *Ergonomics, 15,* 545–554

Johansson, S. (1986). Perceived exertion, heart rate and blood lactate during prolonged exercise on a bicycle ergometer. In G. Borg & D. Ottoson (Eds.), *The perception of exertion in physical work* (pp. 57–68). London: Macmillan.

Jones, N., Sutton, J., Taylor, R., & Toews, C. (1977). Effect of pH on cardiorespiratory and metabolic response to exercise. *Journal of Applied Physiology, 43,* 959–964.

Kostka, C. & Cafarelli, E. (1982). Effect of pH on sensation and vastus lateralis electromyogram during cycle exercise. *Journal of Applied Physiology, 52* (5), 1181–1185.

Lollgen, H., Graham, T., & Sjogaard, G. (1980). Muscle metabolites, force, and perceived exertion bicycling at varying pedal rates. *Medicine and Science in Sports, 12,* 345–351.

McCloskey, D., Gandevia, S., Potter, E., & Colebatch, J. (1983). Muscle sense and effort: motor commands and judgements about muscular contractions. In J. Desmedt (Ed.), *Motor control mechanisms in health and disease.* New York: Raven.

Noble, B., Borg, G., Jacobs, I., Ceci, R., & Kaiser, P. (1983). A category-ratio perceived exertion scale: Relationship to blood and muscle lactates and heart rate. *Medicine and Science in Sports and Exercise, 15,* 523–528.

Pandolf, K. (1983). Advances in the study and application of perceived exertion. *Exercise and Sport Sciences Reviews, 11,* 118–158.

Poulus, A., Docter, H., & Westra, H. (1974). Acid-base balance and subjective feelings of fatigue during physical exercise. *European Journal of Applied Physiology, 33,* 207–213.

Robertson, R. & Metz, K. (1986). Ventilatory precursors for central signals of perceived exertion. In G. Borg & D. Ottoson (Eds.), *The perception of exertion in physical work* (pp. 111–121). London: Macmillan.

Robertson, R., Gillespie, R., McCarthy, J., & Rose, K. (1979). Differentiated perceptions of exertion. Part II. Relationship to local and central physiological responses. *Perceptual and Motor Skills, 49,* 691–697.

Robertson, R., Caspersen, C., Allison, T., Skrinar, G., Abbott, A., & Metz, K. (1982). Differentiated perceptions of exertion and energy cost of young women while carrying loads. *European Journal of Applied Physiology, 49,* 69–78.

Robertson, R., Falkel, J., Drash, A., Swank, A., Metz, K., Spungen, S., & LeBoeuf, J. (1986). Effect of blood pH on peripheral and central signals of perceived exertion. *Medicine and Science in Sports and Exercise, 18* (1), 114–122.

Sargeant, A. & Davies, C. (1973). Perceived exertion during rhythmic exercise involving different muscle mass. *Journal of Human Ergology* (Tokyo), *2*, 3–11.

Sutton, J., Jones, N., & Toews, C. (1981). Effect of pH on muscle glycolysis during exercise. *Clinical Science, 61*, 331–338.

Young, A., Cymerman, A., & Pandolf, K. (1982). Differentiated ratings of perceived exertion are influenced by altitude exposure. *Medicine and Science in Sports, 14*, 223–228.

Author Index

Subject Index